PRAISE FOR

THE HORSE

Holiday Gift List, *Equus* magazine
***Discover* magazine's "What to Read in November" (2015)**
Staff Pick at Changing Hands Bookstore, Phoenix, Arizona
Staff Pick at Kepler's Books, Menlo Park, California
Staff Pick at Eight Cousins, Falmouth, Massachusetts

"Whether you believe that you know horses intimately or only admire them from afar, you will find Wendy Williams's fascinating natural history *The Horse* to be illuminating. Williams is a charming tour guide through the history of the horse-human bond. This book will delight you and deepen your understanding of the ongoing love affair between humans and their equine companions."
—ELIZABETH LETTS, bestselling author of *The Eighty-Dollar Champion*

"Lifelong equestrian enthusiast Williams takes on the topic at full gallop, weaving scientific analysis with cultural and historical anecdotes in this lively, fascinating read."
—GEMMA TARLACH, *Discover*

"[A] fascinating account of the relationship between humans and horses. Not only will horse lovers find *The Horse* a great read, but so will those with just a passing interest in these animals."
—WILLIAM HAGEMAN, *Chicago Tribune*

"An informative and engaging account of an animal that's both familiar and mysterious . . . [*The Horse*] provides an illuminating glimpse at what we know—and what we may someday learn."
—MEGAN McDONOUGH, *The Washington Post*

"Simultaneously a fascinating natural biography; a scientific travelogue . . . ; and a compelling exploration of the bond that unites horses and humans."
—ELEANOR O'HANLON, *BBC Wildlife*

"Engaging, comprehensive . . . Wendy Williams combines a love of horses with a keen interest in natural history . . . An accessible read and a gift for horse lovers."
—LUCY POPESCUE, *The Independent*

"[*The Horse*] takes us on both an intellectual journey and an equine adventure."
—FRAN JURGA, *Equus*

"A stirring journey with a tremendous amount of heart, wonder and scientific detail . . . [*The Horse*] is a book that horse people will read and reference among their friends for years to come." —LORRAINE JACKSON, *Horse Nation*

"*The Horse* will change the way you see equines forever . . . Wendy Williams takes a deep dive into the vastly complex and sometimes tenuous evolutionary journey responsible for the creatures we know and love today."

—NINA FEDRIZZI, Horse Collaborative

"A personable, enlightening account of Williams's work and travels across the globe . . . Reading [her] book will help you appreciate horses' long journey to your barn and pasture." —*NickerNews* blog

"Outstanding and groundbreaking . . . Williams gives us a cohesive, comprehensive, yet comprehensible study of this magnificent animal."

—JEREMY TOWNSEND, *Books for Animal Lovers*

"A compelling journey into the evolutionary history of the horse we know and love today . . . It's detailed but clear, and surprisingly fun in places as well."

—*The Eloquent Equine*

"An ambitious undertaking . . . Reading *The Horse* will no doubt cause you to look at horses (including your own) in a whole new way."

—DALE LEATHERMAN, EquiSearch

"A magnificent natural history of this magnificent animal. Wendy Williams pursues the wild and ancient creatures who put into relief the very particular horse-human relationship we have today. If you get a thrill when a horse thunders by, you must read this book."

—ALEXANDRA HOROWITZ, bestselling author of *Inside of a Dog*

"For every equestrian, here finally is a book that explains in great detail the long journey that has given us the modern horse in all of its magnificent varieties. Wendy Williams's *The Horse* is full of love and firsthand experiences that make the book a most pleasurable and informative read."

—FRANS DE WAAL, author of *Peacemaking Among Primates*

"An enthusiastic history of and appreciation for all things horse . . . Anyone with a love of horses will treasure this book, which provides scholarly yet accessible insight into a beautifully constructed animal that has chosen to domesticate man, just as dogs have." —*Kirkus Reviews*

"Williams's book educates, entertains, and enthralls; it's part scientific discovery, part social commentary, and part history lesson, while always focusing on the relationship between horses and humans." —*Publishers Weekly*

"What a gift. Reading *The Horse*, I felt a kinship with horses and a deep yearning to bond with these marvelous creatures."

—JOHN W. PILLEY, bestselling author of *Chaser*

"Riveting and moving. A beautiful celebration of the deep evolutionary fellowship between horses and people." —DAVID GEORGE HASKELL, author of the Pulitzer Prize finalist *The Forest Unseen*

"I have never owned horses, but reading Wendy Williams's fine new book made me feel as if I've known them all my life. And that, of course, is her point—to explore not just the history of horses but the human fascination with them. She makes her case in clear, compelling prose, warmed throughout by her obvious fondness for the subject."

—THOR HANSON, author of *The Triumph of Seeds*

"*The Horse* is a scientific ode to one of the most charismatic mammals on earth—it is an essential book and a loving exploration of our complicated relationship with the charming animal that we've variously hunted, tamed, and venerated. Readers could not ask for a better guide than Wendy Williams as she takes us to the badlands where mustangs still roam, the bones of their ancestors beneath their hooves."

—BRIAN SWITEK, author of *My Beloved Brontosaurus*

"Noble, intelligent, graceful. These are words we use for the best horses. They apply also to Williams's book. I guarantee that after reading it you will never look at a horse in the same way again."

—JOHN LEWIS-STEMPEL, author of *Meadowland*

ALSO BY WENDY WILLIAMS

Kraken

Cape Wind (with Robert Whitcomb)

Best Bike Paths of New England

Best Bike Paths of the Southwest

The Power Within

WENDY WILLIAMS

THE HORSE

Wendy Williams is a journalist whose work has appeared in *The Boston Globe*, *The Wall Street Journal*, *The New York Times*, and *The Christian Science Monitor*, among many other publications. She is the author of several books, including *Kraken* and *Cape Wind*. She lives in Mashpee, Massachusetts. Follow her on Twitter at @TheNobleHorse.

THE

THE EPIC HISTORY OF OUR NOBLE COMPANION

SCIENTIFIC AMERICAN / FARRAR, STRAUS AND GIROUX
NEW YORK

HORSE

WENDY WILLIAMS

Scientific American / Farrar, Straus and Giroux
18 West 18th Street, New York 10011

Printed in the United States of America
Published in 2015 by Scientific American / Farrar, Straus and Giroux
First paperback edition, 2016

An excerpt from *The Horse* originally appeared, in slightly different form, in *Scientific American*.

The Library of Congress has cataloged the hardcover edition as follows:
Williams, Wendy, 1950–
The horse : the epic history of our noble companion / Wendy Williams. — First edition.
pages cm
Includes index.
ISBN 978-0-374-22440-0 (hardcover) — ISBN 978-0-374-70977-8 (e-book)
1. Horses—History. 2. Horses—Evolution. I. Title.

SF283 .W55 2015
636.1—dc23

2015003860

Paperback ISBN: 978-0-374-53660-2

Designed by Abby Kagan

1 3 5 7 9 10 8 6 4 2

FRONTISPIECE: Tecumseh facing off his rivals, photograph by Greg Auger

For all the horses
all over the world
who have carried me along
on their great adventures

And for a very special person,
Diane Davidson,
a great friend to me
and to the ocean

There is no limit to their treasures;
their land is full of horses
—Isaiah 2:7

CONTENTS

Arctic Ocean
WHITEHORSE, CANADA:
Yukon horse discovery
NORTH AMERICA
GREENFIELD, N.H.:
HorseTenders
SOUTHEASTERN VERMONT:
Former home of Whisper and Gray
SABLE ISLAND:
Home to bands of evolutionarily unique horses
BERINGIA:
Not just a bridge, but a place to call home
POWELL, WYO.: Polecat Bench
ROYAL, NEBR.:
Ashfall
LOS ANGELES, CALIF.:
La Brea Tar Pits
NEW HAVEN, CONN.:
Peabody Museum of Natural History
SOUTHERN CALIFORNIA:
Home of Karen and Lukas
NEW YORK CITY, N.Y.:
American Museum of Natural History
MISSISSIPPI:
Chris Beard discovers 56-million-year-old horse and primate teeth
Pacific Ocean
Atlantic
EQUATOR
SOUTH AMERICA
CHILE:
Charles Darwin discovers horse tooth
0 Miles 3000
0 Kilometers 3000
Scale at the Equator

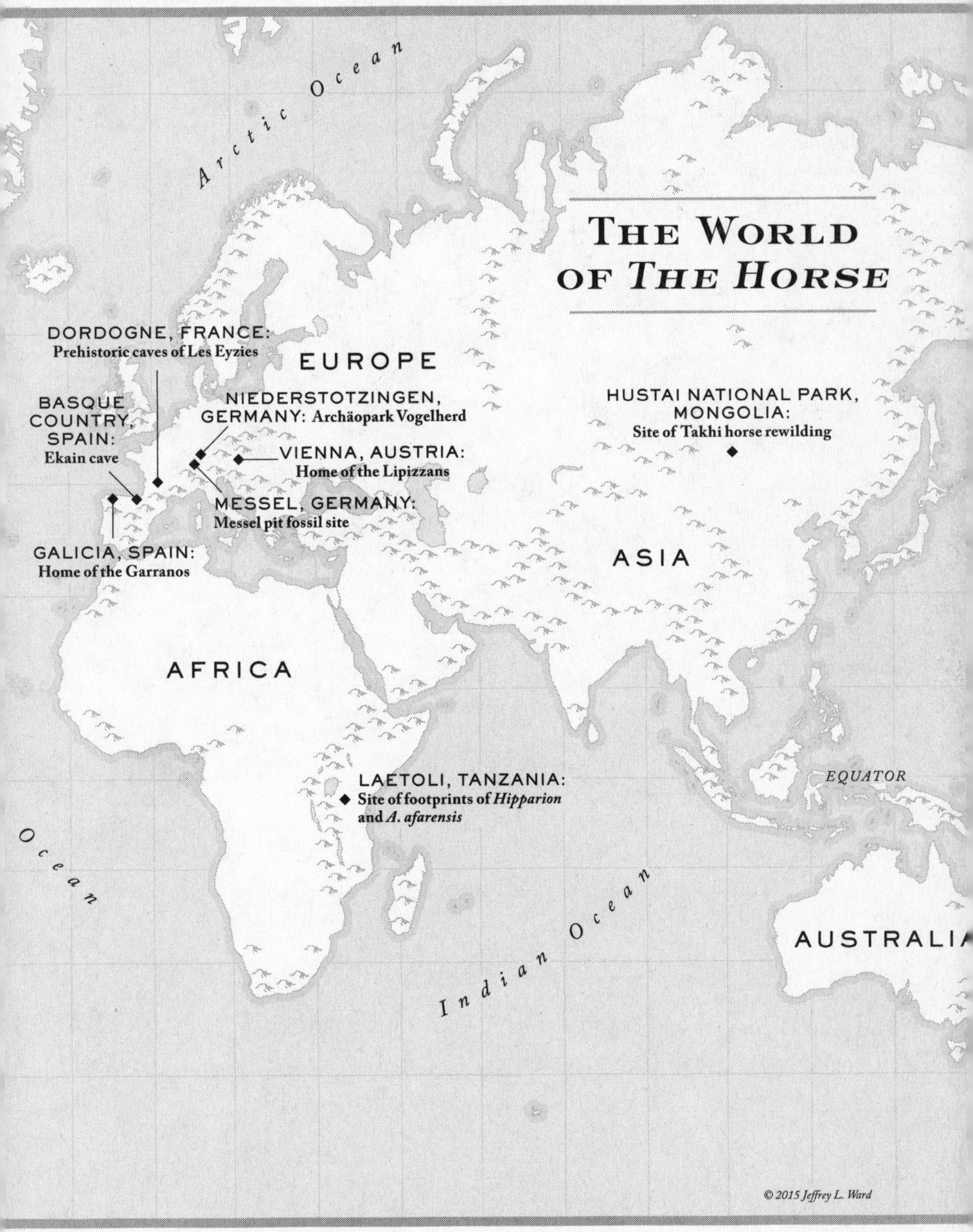

THE WORLD OF *THE HORSE*
Arctic Ocean
EUROPE
DORDOGNE, FRANCE: Prehistoric caves of Les Eyzies
BASQUE COUNTRY, SPAIN: Ekain cave
NIEDERSTOTZINGEN, GERMANY: Archäopark Vogelherd
VIENNA, AUSTRIA: Home of the Lipizzans
MESSEL, GERMANY: Messel pit fossil site
GALICIA, SPAIN: Home of the Garranos
HUSTAI NATIONAL PARK, MONGOLIA: Site of Takhi horse rewilding
ASIA
AFRICA
LAETOLI, TANZANIA: Site of footprints of *Hipparion* and *A. afarensis*
EQUATOR
Ocean
Indian Ocean
AUSTRALIA
© 2015 Jeffrey L. Ward

THE HORSE

PROLOGUE: BACKYARD HORSE

> He trots the air;
> The earth sings when he touches it.
> The basest horn of his hoof is more musical than the pipe of Hermes.
>
> —WILLIAM SHAKESPEARE, *Henry V*

I used to have a little half-Morgan palomino whom I most certainly did not deserve. The horse was a gem, although I was too young and too ignorant to know that at the time. I took him for granted. Mornings I saddled my horse and rode him down our Vermont dirt road to the grade school where I taught piano. There my gelding would stand in his halter, trying desperately to grab bites of grass while the kids out for recess annoyed the poor thing, in a loving sort of way, endlessly.

The kids were overjoyed to have a horse in their school field. I don't believe Whisper shared in their pleasure. I suspect that to him the swarm of excited small children was somewhat akin to a plague of large horseflies—something to be patiently endured. Nevertheless, Whisper was unfailingly tolerant. He was the most polite horse I have ever had.

In the winter, I tied small logs to the pommel of my Western saddle and had him pull them over the snow and through the woods to my house, where they were destined for the woodstove. Having seen such things on television, it seemed to me a romantic thing to do. Whisper, I am certain, did not share my sense of the exotic. I got the BTUs from those logs. He got the hard work. He was much put upon, but never once did he kick me or otherwise complain, although he certainly had a right to. I wish I could live my own life with as much dignified fortitude.

Some people claim that horses, by accepting such behavior in humans, show their lack of intelligence. I don't believe that. Whisper was a persistently pragmatic fellow who, if he couldn't get what he wanted at the front door, would find a way to go around to the back. He was, by necessity, highly skilled at getting his own needs met. He was a mastermind, an equine Einstein, a street genius, and a determined survivor, as are so many horses cared for by humans who, like me, are basically clueless about their horses' inner lives.

In my small Vermont hillside barn I had only two stalls and no running water. I usually carried the water buckets down the hill from the house's outdoor spigot. This involved quite a bit of labor on my part. Once, rather than carry the full buckets all that way, I brought Whisper and his half-Percheron buddy Gray up the hill to drink from a bucket that sat under the outdoor faucet. I thought I'd had a brilliant flash of insight.

But allowing Whisper in on the secret of where the water came from would turn out to be a serious mistake. Months later, I was out having way too much fun and returned quite late to feed and water. I felt uneasy about this, but, since I was only twenty-two, not all that uneasy. The horses wouldn't die if their routine were delayed. What was the big deal?

But as any barn hand knows, horses have a different point of view. The stamping and weaving and snorting and stall-chewing will start almost immediately after the appointed feeding time. Then the panic will escalate. Horses get anxious when their expectations are not met.

Some horses resign themselves to a late feeding. Others do something about it.

When I pulled into my driveway that night, a huge puddle filled my yard. I found my house's outdoor spigot on full tilt. This seemed inept, even for me. Perhaps a friend had been by and watered for me. But I found no note. When I checked the barn, the water buckets were empty. The mystery heightened.

After filling the buckets, I walked up from the barn, noting the winter weather. The sky was clear; my conscience—not so much. Callow and inexperienced as I was, I nevertheless realized I had failed to live up to my side of the horse-human partnership.

One day a week later, I rose a little late. The thermometer registered minus ten. So perfect in the summertime, Vermont sure can be cold in the winter, I whined to myself. I needed, I was sure, a hot cup of coffee before I tackled the barn. Maybe even two.

That was my point of view.

Whisper felt differently. I looked out the kitchen window down the hill at the barn. I slowly sipped my coffee. Over the pasture fence came that little golden-coated horse. His knees were tucked up into his chest like a champion athlete taking on a Grand Prix course. It was such a big and perfectly executed jump that I was taken aback. I hadn't known he could jump at all.

A horse of hidden talents, I thought.

Once clear of the fence (his soaring hoofs never touched it), my Morgan adopted a pleasant jog and headed himself straight to the water spigot.

Wham, wham, wham.

The water was on.

Hoofs, I learned that day, have a variety of uses.

Next Whisper stretched his lips out to make a kind of cup. This was also something I had never known a horse could do. He let the water cascade from the spigot into his mouth. Horses, it turns out, have highly sensitive and tactile lips that are much more agile than our own.

Having gotten what he wanted, Whisper strolled back down to the barn and waited for breakfast.

Horses can be quite ingenious when they're motivated, and when it comes to water, their motivation is intense. But my Morgan's cognitive genius was not limited to water witching. Clearly, Whisper could solve many kinds of problems—getting around an electric fence, for example, or opening his stall door. When he had specific goals in mind, he was a high achiever.

This was all very nice for Whisper, but I didn't necessarily want him meeting his own needs by running all over the countryside. My neighbor with a splendid front lawn of luscious grass had already told me that horses frightened him.

Of course, motivation varies from individual to individual. Some horses are better than others at perfecting basic survival skills. My workhorse, Gray, rarely innovated. After Whisper jumped the fence and drank his fill from my water spigot, I threw on my jacket and made my way down to the barn. Gray was standing stolidly in his stall, waiting for me. Whisper's stall door stood wide open. Looking in, I saw the problem: both water buckets were full (I had at least learned that much), but the water was frozen solid. The workhorse had expected me to solve the problem for him. The Morgan had solved it for himself.

Just how clever was my half-ton imp of a horse? I wondered. And how did Whisper's inventive mind compare to Gray's? I designed an experiment. I left several apples just out of reach while both horses stood in their stalls with the latches fastened. Both could reach their heads over the half doors but—theoretically—would remain where they were until some human lifted the latch.

I stood and watched. Both horses eyed the apples. Neither took decisive action. Then I left the barn and acted as though I was heading up to the house. But once out of their sight, I stopped and spied on both through a barn window. With no hesitation at all, Whisper reached over and shifted the latch with his all-too-nimble lips. He pushed open the stall door, walked out, and enjoyed both apples. Gray just watched. Now I knew for sure what I had earlier just suspected: Whisper was free to come and go from his stall whenever he chose.

And even more important, not only was Whisper clever enough to get the apples, but he was also clever enough to hide his knowledge from me. What goes on in the world of horses when no one is looking is often different from what goes on when we're around. Horses can be quite secretive. They are well aware that rules made by humans can be broken when the humans are no longer in the barn.

I had a lot of questions. How did Whisper formulate his plans? Did he have any "plan" at all? Was he conscious? Could he think? Did he have what scientists call "theory of mind"—an ability that, among other things, lets us discern what others may be thinking—as his sneaking of the apples implies? Many of us, myself included, were raised to think of horses as simple automatons that we human beings, as masters, must dominate, direct, and control. Indeed, we often work with horses in terms of very simplistic behaviorism—reward and punishment.

But horses are much more complicated than that. After 56 million years of evolution, after surviving planetary deluges and repeated ice ages, why wouldn't they be? Whisper taught me how very wrong my early education had been. He was far from a machine; he was a living being with ideas of his own and with obvious decision-making abilities. But how did he make those decisions? What were his criteria? Did he enjoy some sense of himself? It certainly seemed so.

I don't mean to imply that horses are just like humans. I'm fairly sure that Whisper did not think to himself: "Mmmm . . . apples . . . I'll just wait until she leaves. Then I'll lift the latch and head right over. No need to worry about that old workhorse. He's not smart enough to get there first."

Instead, he had his own integrity, a unique outlook based on having four legs, a pretty good brain, a liking for grass, a need for water, and a dislike of novelty. He certainly had some kind of organized approach to problem solving, coupled with strong motivation and lots of curiosity. Of course, food can be a terrific motivator. Anyone who has ever watched a pastured horse reach persistently for the greener grass that's perpetually on the wrong side of the fence will know that.

The more I thought about Whisper, the longer my list of questions grew. Where do horses come from? Why do they have hoofs and not, like us, fingers and toes? Why are they willing to share their lives with us? What biological roots, laid down in deep time, created the foundation for our mutual partnership? How does this shared ancestry allow us to understand each other? Dogs and humans can read each other's body language. Can horses read each other's body language? Can they read ours? Do they even bother to try? My list of questions was endless. The more I learned, it seemed, the more I wanted to know.

We are, like horses, children of the savanna, offspring of the wind and the sun and the pelting rain. This is more than just a romantic idea. In the last few decades, science has corroborated the romance with a growing body of research. This science is helping inform us about how to treat a horse well in the modern world, about the hidden emotional lives of horses, about whether they'd be happier living "free" out on the open plain or whether they prefer the safe haven of a barn and regular mealtimes, and even about their social and cognitive requirements.

In the days when I looked after Whisper and Gray, I never thought about such matters. My main objective was to figure out how to keep the Daring Duo safely in their stalls and in their pasture and off my frightened neighbor's delicious lawn and out of the grain bin, which Whisper had also learned how to open.

I wasn't alone in failing to consider the horse's basic needs, other than food and water. As I'll explain, horses were integral to our existence long before we had advanced culture. You might even say they *gave* us human civilization. Nevertheless, although they've been domesticated for more than six thousand years (no one knows exactly how long), it's only recently that we have come to see them as sentient beings with finely nuanced minds. What took us so long?

The Horse is a scientific travelogue, a biography of the horse, and a worldwide investigation into the bond that unites horses and humans. By visiting and talking with scores of scientists from all over the world, in places like Mongolia and Galicia (in the northwest corner of Spain), with archaeologists studying prehistoric sites in France and the Basque country, with paleontologists in Wyoming and Germany and even downtown Los Angeles, I uncover the shared journey of horses and humans over time, examine our biological affinities and differences, and discuss the future of horses in a world filled with people.

This is also my belated ode to Whisper, and to his buddy Gray, and to all the other horses I've encountered in my life who have so kindly and patiently carried me over the Rocky Mountains and into the Sahara Desert, who have joined me in my meandering along the overgrown dirt roads of Vermont, who have walked me safely past crocodiles and hippos and grizzly bears—all of whom have taught me a whole lot more than the fact that water in buckets freezes in the wintertime.

Last but not least, *The Horse* is an ode to all the horses who have for many thousands of years helped make human life so much better. As George Gaylord Simpson, a renowned expert in horse evolution, once wrote, "From horses we may learn not only about the horse itself but also about animals in general, indeed about ourselves and about life as a whole."

1

WATCHING WILD HORSES

> There is no doubt that horses will exist as long as the human race, and that is well, for we still have so much to learn about them.
>
> —C. WILLIAM BEEBE, naturalist

Sometime around thirty-five thousand years ago, when much of Europe was locked up in sheets of ice that pulsated sluggishly over the land like frozen heartbeats, an unknown artist acquired a bit of mammoth ivory. Perhaps he found the ivory lying on the ground. Or maybe a group of hunters brought it to him as an offering.

This mysterious craftsman possessed phenomenal skill. Wielding with great precision a set of exquisitely honed stone tools, he began carving a masterpiece. A magnificently arched stallion's neck appeared, breathtaking in its extraordinary combination of muscular potency and simple natural grace.

The earliest example of an archetype that has since then appeared in art worldwide, this horse embodies the essence of majesty. He is the supreme example of Platonic form, "an abstraction of the graceful essence of the horse," in the words of the anthropologist Ian Tattersall, or, more simply, the *rasa* of horse, to use a Sanskrit term. The curvaceous line of his head and neck flows smoothly into his withers and backline, creating an elegant S curve that finishes just below the hindquarters. The head, slightly cocked, gives the animal an air of fortitude and deep contemplation.

When we see him, we love him. And we recognize him: This sculpture could have been carved only yesterday. Across thirty-five millennia, you can almost hear him snort and see him toss his head, warning encroaching stallions to take care. Called "esthetically perfect" by his current curator, Harald Floss of Germany's University of Tübingen, this two-inch-long marvel, standing only about an inch high, is known as the Vogelherd horse, in honor of the cave in southern Germany in which it was discovered.

The 35,000-year-old Vogelherd horse, the oldest known horse sculpture, carved of mammoth ivory (Museum der Universität Tübingen)

The carving provides evidence that the emotional bond between horses and humans began long, long ago—tens of thousands of years before human civilization began, well before horses became domesticated, well before we kept horses in our barns and in our fields to be used as tools. We have no idea who created this tour de force, but we do know one thing: this ivory carver spent a lot of time watching wild horses, studying their social interactions and learning their body language. He carved his subject confidently, with a sure hand.

We also know that the artist was a member of the first group of thoroughly modern humans to create a substantial presence in

Europe. These people, Aurignacians, revered not just horses but many animals. Their art is exquisite—but it's so much more than that. It's a scientifically valuable body of evidence that provides us with precious data, including a record of the wildlife with which early humans shared Ice Age Europe's river valleys, marshlands, and open plains. This record consists of an endless array of painted caves, countless bas-relief sculptures, sketches and etchings, and many, many more carvings—all of which depict, sometimes in great detail, the strange animals like woolly rhinoceroses living in Pleistocene Europe.

Some of these creations are of impressively high quality—and yet, the art is far from rare. In fact, it's curiously omnipresent. Archaeological sites containing art from this time have been found in western Spain, in Italy, and in France, and all the way east into Russia. A modern admirer could easily set aside a whole summer to study them and still have seen only a small portion. Nevertheless, common as this art is, its mere existence is almost miraculous: Aurignacian art appears in the European archaeological record seemingly quite suddenly, as though a genie waved a hand and humans became creative. There are no obvious precursors, no clear antecedents that show any kind of learning curve. Of course, archaeologists say, this could not be literally true. There must have been some learning period, complete with an upward-moving arc of acquired skills, but, as of now, almost no evidence of this arc has been discovered.

The phenomenon is so remarkable that some researchers once suggested that the *Homo sapiens* brain, already around for well over a hundred thousand years, may have undergone a sudden neurological advance—some shift in the human psyche that brought about the creative impulse. That theory is no longer in vogue, but it is clear that something monumental had occurred. Otherwise, scientists are at a loss to explain the ivory carver's tiny talisman.

The Vogelherd horse, caught in the act of behaving with such supreme hauteur, is so much more than a simple symbol—he's a living animal, frozen in a specific instant in time. He is about to strike out with a forefoot, or perhaps about to sidle up to a mare. He is the modern Friesian stallion pacing anxiously in our pastures, just about to shake his head, or the American mustang* running free on the open plains, about to pose against some red-rock cliffs, or the accomplished dressage horse about to execute a perfect *piaffe*, that beloved classical movement that shows off a horse's contained energy and flowing grace.

All of this begs the question: Why? Why did the artist care so much about a *horse*? Was this a religious icon? Was it tradable currency? Did it confer a stallion's energy on its human possessor? Or was it perhaps not important at all, but just a toy made one winter afternoon to entertain the kids?

Whatever its purpose, this stallion was not put on a pedestal and simply admired. He was handled. A lot. The artist carved tiny lines into the horse's back, and the lines are now well-worn by having been touched many times by human hands.

The answers to our questions may be forever elusive, but we do know one thing. We share with the ancient artist a powerful emotional response: we today are just as mesmerized by horses as were people thirty-five thousand years ago. Even today, separated as we are from the natural world, we yearn for contact with horses. Just ask any mounted policeman.

Although the ancient carving is shrouded in mystery, he had plenty of company. For the next twenty thousand years, until the ice finally melted and Europe entered our present warm period, artists created horses in whatever medium they favored—ivory, antler, wood, stone, paint.

Horses are the stars of Ice Age art. Indeed, horses are the most frequently represented animal in the twenty-thousand-year period that preceded the advent of farming and what we call civilization. At Abri de Cap Blanc in France, a fifteen-thousand-year-old rock overhang under which people lived, a nearly life-size bas-relief of horses was carved into the rock wall that served as a backdrop for day-to-day family life. When I visited this site, the stone carving reminded me of kitchen art—something to ponder while you stir the soup—yet the

* "Mustang" in this book does not denote a breed, but a horse of the American West born free on the open range, as opposed to a domestic horse, which I define as a horse born in captivity.

Cap Blanc horses are as vivid as any created by Leonardo da Vinci. They seem to come alive and jump out of the rock when light flickers over them.

Hundreds of miles west of Cap Blanc, in the caves of the Spanish north coast, sensitively drawn ponies frolic on the walls with joy and abandon. Thousands of miles east, in Russia's Ural Mountains, horses sketched in red ocher grace the walls of Kapova Cave. On the walls of Chauvet Cave in southern France, painted horses stand in small groups, watching the wildlife around them, including lions prowling nearby. Some Chauvet horses graze while others keep watch. Elsewhere in the cave, a timid horse peeks out from behind a rock. What is he afraid of? The hunting lions? A powerful stallion?

Ice Age artists seemed to know everything about horses. Until Leonardo came along and actually studied the horse's anatomy, no other artists equaled these Pleistocene virtuosos in their portrayal of what it was like to be a horse. To me, these first-known, highly accomplished artists are also the world's first animal behaviorists. They must have spent hours and days and months and years just watching. They understood horses' facial expressions, how their nostrils flared when they were frightened, how their ears betrayed their inner emotions, how they sometimes stood together in small bands, and how, sometimes, they would wander alone and seem rather forlorn. From this art, we know that long before horses became our tools, long before the bit and the bridle were invented, we *Homo sapiens* adored watching wild horses.

Sadly, though, in the modern world, this has become something of a lost art. While we enjoy *seeing* free-roaming horses, few of us sit quietly and study them in depth. Consequently, we suffer from lack of context. We see what the horse is doing, but we don't always know why he's doing it. We know little about how horses *really* behave when they're out of our sight. We see horses standing in our barns and pastures and mistakenly assume that what we see is the *essence* of "horse." I've always thought this rather strange.

On the other hand, ethologists study the behavior of lions in the wild, of birds, of monkeys, of whales, and of elephants. Their research has enriched our view of what it means to be part of the living universe, so that we now understand that we all fit into a finely woven web, and that this web, in a reasonably healthy state, is central to our own well-being. We may be top dog when it comes to creating an electronic society, but other animals have talents in other areas that far exceed ours.

This revolution in our understanding of the natural behavior of animals was brought into the public spotlight in the 1960s by authors such as the Nobel Prize–winner Konrad Lorenz, whose bestselling books included *King Solomon's Ring* and *On Aggression*. Lorenz was particularly well-known for establishing scientifically the importance of attachment in the lives of animals. He emphasized that studies of animals in laboratory settings did not reveal the true nature of various species. To understand that, he wrote, animals must be observed in the context in which they naturally live.

His books caused a worldwide transformation in our thinking about wildlife. Young scientists from many different nations set up research sites in remote parts of the world and methodically recorded the behavior of the animals they watched. For example, for more than forty years, Jane Goodall and her team have studied chimpanzees at Gombe Stream National Park in Tanzania. When Goodall began her work, she shocked—"shocked" is not too strong a word—the world by reporting that primates commonly fashioned and used tools. So much for the formerly unshakable status of humans as the only toolmakers on Planet Earth. At about the same time, in the 1960s, Roger Payne and Scott McVay studied the behavior of humpback whales and found that they communicate with each other by singing what Payne called "rivers" of sound. So much for the status of humans as the only beings with sophisticated communication systems. Crows are adept at creative problem solving. Octopuses use their arms to open jars, to build complicated rock shelters, even to carry seashells in case they need emergency housing. Elephants use teamwork to protect family members. Bats echolocate. Bees have swarm intelligence.

But what about horses? What are *their* special powers? How much has modern ethology learned about the natural behavior of horses? Not much, it turns out. Why? If our fascination with the details of horse behavior stretches back at least thirty-five thousand years, as the evidence shows us, why have horses been left out of this scientific reformation? Equine scientists have studied the best way to train show horses, the best way to feed racehorses, the best way to heal the delicate bones in a lame horse's feet. But the natural behavior of horses was rarely considered to be of scientific interest. Only a handful of ethologists had watched wild horses in a methodical manner. And of the studies that were done, very few were long-term projects akin to those of Jane Goodall.

That's beginning to change.

Jason Ransom and I were talking about this one July evening in Cody, Wyoming, a gateway town near Yellowstone National Park, founded by Buffalo Bill Cody and his Wild West showmen more than a century ago. Ransom, an equine ethologist, and I had met a year earlier at an international conference held in Vienna that was attended by scientists who study free-roaming equids—horses, zebras, onagers, wild asses, and donkeys—at sites located around the world. Ransom invited me to come and meet some of his study subjects—several populations of wild horses who roam regions of Wyoming and Montana. He had followed their behavior over a period of five years and discovered some behaviors that upended many long-held myths about how horses bond with and interact with each other.

Meeting up in Wyoming, we spent several days watching Ransom's study subjects and watching the people who came from around the world just to watch the horses. Like our Ice Age ancestors, these people sat for hours, enjoying the scene. Small groups talked among themselves, discussing the horses and interpreting their actions. Some people even camped out, so they could observe the horses for a complete twenty-four-hour period. It made for an entertaining time, and I could easily imagine a similar scene tens of thousands of years earlier: people relaxing in the summer sunshine and discussing what the horses were up to.

That particular July evening, after a great day of wild horse watching, Ransom and I were leafing through one of several books of Ice Age art that I had brought. As we looked at the reproductions of

Ice Age prancing ponies painted on the walls of many different caves in France and Spain, we talked about how many of the complex behaviors shown on those prehistoric cave walls were behaviors we had seen in real life just hours earlier.

We discussed the power of horses over the ancient human mind and compared it to the power of horses over the *modern* human mind. Horses and humans, we realized, have so much in common: we are both the result of tens of millions of years of planetary upheavals, of the ebb and flow of plant life, of rising mountains and shifting ocean currents. Because of this common evolutionary heritage, we are drawn to horses in a way that's rudimentary, elemental, even atavistic. Consider the tantalizing story of Nadia, an autistic savant who, at the age of three, broke out of her shell by suddenly—spontaneously, without any training at all—drawing spectacular galloping horses, horses with flowing manes and tails, horses created from memory but perfectly, sublimely depicted in correct proportion. Nadia could have chosen to draw any number of animals, but what drew her attention was horses. Perhaps, Ransom and I thought, a fascination for horses is somehow encoded in our genes. When we see horses running on an open plain, we imagine ourselves doing the same thing. Even when people are separated from the natural world, even when they spend most of their lives in twenty-first-century cities, horses still speak to something essential in us, just as they spoke to something essential for the carver of the Vogelherd horse.

"Most people today are not at all familiar with what it is to be a horse," Ransom said, "but they can still see a picture of a horse and love it. What is that? What is the factor that connects us?"

That our love affair with horses has been going on for tens of thousands of years, Ransom and I decided, speaks volumes. And yet, we modern people misunderstand horses in some important ways. Since the days when I looked after Whisper and Gray, I have had many horses and happily spent more than my fair share of time in the saddle. I thought I knew a lot about how they behaved. But under Ransom's tutelage, I realized that I knew almost nothing about them, save for how they behave in a barn or paddock.

By watching wild horses*—horses born away from human contact, as opposed to domesticated horses, raised around humans—I learned that horses are exquisitely sophisticated animals, capable of all kinds of unexpected interactions. And I learned that the act of watching is so much more interesting if you know the backstories of the individual animals you're looking at. You'll learn that various horses often set their own agendas, and you'll slowly come to understand those agendas. That's when things get intriguing, because each horse has his or her own personality. As I learned with Whisper and Gray, one horse may take bold action to solve a problem while another may choose a more passive course. But that doesn't mean that the passive course is any less goal-directed.

Horses are different from many of the other ungulates—hoofed mammals—who populate everything from savannas and grasslands to forests and rocky outcroppings around the world. Ungulates are common. Cows and goats and sheep, bison and deer and moose are all ungulates. However, unlike many ungulates, who seek safety in numbers and who roam the plains in large groups, horses form intimate social bonds, just as elephants do. With horses, though, those bonds, while strong, are also quite fluid. As with humans, friendships come and go, foals grow up and depart to live elsewhere, and male-female relationships sometimes work out and sometimes don't.

These close bonds are essential to the horse's psyche. Lacking the opportunity to form such attachments, the natural horse becomes a different animal. His social world is his raison d'être, the foundation of his existence and the reason why he does many of the things he does. After all, in the natural world in which horses evolved, solitary horses usually didn't survive. Nevertheless, contrary to popular belief, science has discovered that they are not "herd" animals. Instead of seeking safety in large numbers, horses live year-round in small groups called bands. Membership in these bands, which may consist of as few as three horses or as many as ten or so, is just as fluid as are the individual

* "Domestic" and "wild" are not, in this book, used as scientific terms, for reasons we'll consider in much greater depth in future chapters.

bonds, but there's usually a central core of closely allied mares and their young offspring.

Like humans, horses in a band are notorious squabblers. Also like humans, band members fail to thrive without friends and family. These attachments are essential. Despite what you see in Hollywood movies, horses, unlike cattle or bison, rarely "stampede" en masse. If several bands of horses grazing in the same area are frightened by something, the bands are likely to head off in all different directions. Their various flight trajectories may look more like spokes in a wheel. This tendency to scatter, if given an opportunity, is one of their survival strategies.

Band members are not simply group animals with gang-like mentalities. As Ransom and other equine ethologists have found, individual bonds within bands may be more important than group identity, just as with us. These bonds are sometimes based on family ties, but often they are just based on individual preference.

When you watch wild horses and you know their life histories, it's like following a soap opera. There's a constant undercurrent of arguing, of jockeying for position and power, of battling over personal space, of loyalty and betrayal. The show never lets up. Alliances are made and broken. Underlings often defy power. Sometimes a horse's great patience is rewarded and he gets what he wants. Sometimes it isn't and he doesn't.

The spectacle is positively Shakespearean. To understand the script fully, you have to pay close attention: like kings and princes, politicians and chimpanzees, some horses act one way in public and then behave quite differently when no one is looking, just as Whisper did.

Earlier that sweltering July day, Ransom and I had watched one of his favorite stallions, Tecumseh, a pinto who roamed a region known as McCulloch Peaks. Tecumseh had been presiding over a stallion squealing contest. As we watched, the bickering got out of hand. Arching his neck, poised for a fight, he looked for all the world like the modern avatar of the Vogelherd horse. Males of many different species like to show their stuff— think of the glory of the peacock's spread tail feathers—but stallions are expert at the art of exuding masculine glamour. They are true drama queens.

As we watched, Tecumseh's whole body wound itself into a warning: Get away from me or you'll be sorry. It wasn't hard to see what he was getting all huffy about. A gang of four pushy boys—too old to be allowed by the mares to hang around the foals and too young to attract mares themselves—had edged up on Tecumseh's personal space. They reminded me of a group of awkward teenage boys sauntering down a city street.

This gang was getting too assertive with Tecumseh and failing to respect the keep-your-distance rule to which horses generally adhere. Even worse, the boys were also implying that they'd like to improve their acquaintance with the mares with whom Tecumseh was then keeping company.

Tecumseh was having none of that. He stared them down. He raised his head and coiled his hindquarters, as if preparing to chase them off. He lifted one foreleg straight up in front of him—then pounded the ground. After he'd shown just how powerful that foreleg was, he lifted and pounded with the other foreleg.

Tecumseh facing off his rivals (Greg Auger)

The four marauders, too young to challenge the old man, walked off to sniff a dung pile.

They had a lot to learn. When stallions fuss with each other the scene rarely heats up into an all-out battle, but you never quite know just how things will play out. In the Pryor Mountains, another region where horses roam freely, Ransom and I saw one stallion, for a reason that was not at all obvious, start to chase a second stallion who was standing quite a distance away. Other stallions were closer to this aggressive male, but he ignored them. Instead, he trotted over to this particular far-off stallion and, snorting and screaming, drove him into a small copse of trees at the far end of the meadow. We lost sight of them. Then they emerged from the trees and galloped across the meadow, creating a general uproar. One chasing the other, they ran to where several other bands were minding their own business. Stallion Number One stretched his head out toward his enemy. He looked like a snake. He bared his teeth. He meant business. Then the two stallions, one fleeing and one chasing, ran too close to a small ridge. This was a poor decision.

Driving away the enemy (Greg Auger)

Beneath this ridge grazed a band of mares accompanied by yet a third stallion. Stallion Number Three, dubbed Duke by researchers, stormed straight up the ridge. Stallion Number One, the initial aggressor, had met his match. Duke, large and well muscled, was full of attitude. A great deal of head-tossing and snorting followed as the first

Duke confronts the marauding stallion. (Greg Auger)

Duke overwhelms his challenger. (Greg Auger)

stallion tried to hold his ground, but all you had to do was glance at the pair to see that he never had a chance. There was no question about who would back down first.

Duke was clearly Lord of All He Surveyed. Stallion Number One skulked away. Stallion Number Two, the one who had been chased all across the meadow, was nowhere to be seen. As the curtain fell on this act, Duke held center stage, displaying his royally arched neck for a few seconds before calmly returning to his grazing.

What had brought on this chaos? Ransom wasn't sure. Adult horses rarely exert themselves in summertime heat unless they have to, but this time thundering hoofs had raged across the whole meadow. As we watched the horses, we noticed that the mares paid little overt attention to the male shenanigans. In fact, in my time watching wild horses, I have never seen mares react to the hostile antics of stallions—as long as the boys kept their problems to themselves.

"That's usually the case," Ransom said when I asked about this. There are some times, he said, when mares will add their opinions to a dispute between stallions, but those times are very, very rare.

In my childhood books I often read about mares huddling together and breathlessly waiting to see the outcome of the stallions' battles, but this is not at all what happens. Mares usually ignore the males' conflicts. This makes sense. After all, if mares stopped eating every time two stallions had a stare-down, the mares would starve to death.

For his doctoral dissertation, Ransom, with the help of several assistants, including the local horse expert Phyllis Preator, recorded the behavior of individual horses living in three different regions in Wyoming and Colorado. That research generated a lot of data that, together with work done earlier on the same horses by other scientists, has created a rich long-term record of the intimate social lives of individual wild horses, one of the few such records anywhere in the world. The data is so thorough that Ransom can sometimes find the birth dates of horses who were, by the time he began his work, already entering old age. He knew where some of these horses had spent most of their lives, knew when some had moved from one area to another, knew when they joined up with specific horses and how long they stayed with their companions before moving on.

Recent ethological research has finally begun to reveal the depth of horses' emotions, but the idea, scientifically speaking, that horses *have* emotions is nothing new. Charles Darwin wrote about horses (and many other animals) in his 1872 masterwork *The Expression of the Emotions in Man and Animals*, which proposed that human emotional expressions are innate and universal, and are shared by a variety of animals. Considered a founding text for the science of ethology, Darwin's book explained that certain basic emotions—anger and fear and disgust, for example—evolved as survival mechanisms early in the history of life. For example, in *The Expression of the Emotions*, Darwin compared his own startle response at the approach of a large snake to the startle response of a frightened horse. Shared emotions, Darwin declared, help us understand the emotions of other species. "Everyone recognizes the vicious appearance which the drawing back of the ears gives to a horse," he wrote. That was certainly true in the Pryor Mountains that day, when Ransom and I watched the snaking stallion with his flattened ears. Neither of us was about to get in the marauder's way.

For Darwin, basic emotions were a universal language, a kind of innate lingua franca common to "man and the lower animals," as he put it. They are survival strategies that, Darwin believed, could be studied methodically across species. Darwin's groundbreaking book set the study of animal behavior and emotions on a solid foundation. It said, scientifically: We are all in this together.

The Expression of the Emotions and Darwin's subsequent writings spawned all sorts of scientific studies of animal behavior, but horses for the most part were left out of this paradigm shift, despite Darwin's discussion of horse behavior in *The Expression*. Perhaps this was just a case of familiarity breeding contempt: we assumed we already knew all we needed to know about horses because domesticated horses were part of our everyday lives.

Now that ethological research principles have begun to be applied to the study of wild horses, we have learned how little we know. Thankfully, researchers are upending many deeply embedded myths. For example, a recent National Academy of Sciences report declared that "a harem, also known as a band, consists of a dominant stallion, subordinate adult males and females, and offspring." Most of us have been taught this and at first glance, it would seem to be true. What we notice when watching wild horses is the uproar created by the stallions.

But research by Ransom and others has shown that this male-centric view is wrong. Far from subordinate, mares frequently initiate the band's activities. The stallions are quite often little more than hangers-on. Ransom was once watching a band of mares who stopped grazing and began heading for water. The stallion didn't notice. When he looked up and saw his female companions leaving, he panicked.

"He started running after them," Ransom told me. "He was like a little boy calling out 'Hey, where's everybody going?'"

The mares ignored him. Whether the stallion caught up or not didn't appear to concern them. Mares also sometimes have stallion preferences. They resist males they don't like with surprising persistence, even when that male has established himself as the band's stallion. The ecologist Joel Berger studied the behavior of two unrelated mares who had spent several years together. The pair joined a band which was then taken over by a new stallion, who tried to assert himself. The two mares refused to accept his attentions. For three days "during numerous forced copulation attempts toward both by this stallion, the females reciprocally aided each other (thirteen times and eighteen times respectively) by kicking and biting the stallion as he persistently and aggressively" tried to mate, Berger wrote in *Wild Horses of the Great Basin.* It's long been known that female elephants cooperate, but before ethologists began systematically studying free-roaming horses, few people suspected that cooperating mares were capable not only of waging such a fight—but of winning it. Given the truth about mares, "harem" seems like such an old-fashioned word.

The biologist John Turner found much the same thing when he studied horses living in high chaparral country on the California-Nevada border. Turner's ongoing research, which has lasted at this writing for almost thirty years, has revealed many occasions when mares were not subordinate to stallions, particularly when one band stallion was driven away by a new stallion. The behavior of these mares was often subtle and indirect, he told me, so that unless observers pay close attention over many years, they might miss the fact that the mares *do* often get their own way.

"Sometimes a mare resisting the change behaves in such a way that the new stallion just lets the old stallion come and take her away," he told me. For the new stallion, coveting these resistant mares may just not be worth the trouble. "It's easy to anthropomorphize some of this," he said, "but sometimes, that's the way it is. Horses do a lot of the same things that we do."

At the conference in Vienna where I met Ransom I also met the Spanish equine ethologist Laura Lagos, who, with the wildlife biologist Felipe Bárcena, studies the behavior of an unusual type of free-roaming horse called a Garrano. Lagos invited me to come visit her study site in Galicia, in northwestern Spain. In this rugged region, shared by horses and wolves and humans for millennia, Lagos and Bárcena have notated the behavior of these free-roaming horses for years, just as Ransom and his team have done in Wyoming and Colorado. The scientists have come to admire the horses' tough, stubborn natures.

Garranos, possibly descended from the horses painted by the Ice Age artists, live rough, tough lives. In the American West, free-roaming horses have few predators, but Garranos must defend themselves from relentless wolf packs. They must be able to thrive in a challenging North Atlantic climate and must live on a dreadful diet of gorse. Sometimes called the Plant from Hell, gorse has sharp thorns in lieu of leaves and can grow waist high. Walking through it without long pants is akin to medieval bloodletting, yet Garranos love this stuff. Many have thick handlebar mustaches that may have evolved to protect their sensitive lips from nasty gashes.

In the course of their study, Lagos and Bárcena cataloged the behavior of a pair of mares in one band who were strongly bonded with each other and who often stood just a bit apart from the rest of the band. At breeding time, the mares went together to visit the stallion of another band. Lagos watched one of the mares consort with this stallion rather than with the stallion from her own band. Then the mares returned to their original group. When the second mare was ready to breed, the duo again deserted their original band and its stallion to consort with the other stallion. Then, again, they returned to their original group. This was not an anomaly. She saw the same pair do the same thing the following year. "They prefer their own territory, but the stallion of the other band," she told me.

The researchers Katherine Houpt and Ronald Keiper, who followed the behavior of several horse bands including the horses of the North American Atlantic coast's Assateague Island, have also found that "the stallions were neither the dominant nor the most aggressive animals . . . and were subordinate to some mares."

I suspect that the myth of stallion dominance has persisted for so long because stallion behavior makes for some pretty enthralling theater. The stallions puff themselves up and snort and squeal, and then, if the battle proceeds, rear up and clash in a frenzy of bellicosity. By contrast, mares going about their business of grazing and raising foals lack the pizzazz factor.

The prevalence of our belief that stallions dominate a band may be due to the hierarchical structure of our own culture, suggests the British researcher Deborah Goodwin. She believes that our own emphasis on dominance has caused us to view relationships among horses with blinders on.

The "blinder factor" may be why we often fail to grasp the flexibility of natural horse behavior. Traditionally, instead of thinking about the relationship between a horse and a human as a partnership, we have thought of it in anthropocentric terms: we dominate and horses submit, according to what we assume is the natural order of things. End of story. Since we don't look closely enough, we misunderstand. We miss the fact that the social lives of mares may be rather complicated. One mare may dominate a second mare, and the second mare may dominate a third mare—but the third mare may dominate the first mare.

Moreover, it turns out that mares don't need to have huge fights to get what they want. Instead, they use the technique of patience.

For example, Ransom believes that only about half the foals in the bands he studied are sired by the band's closely associated stallion. This finding flies in the face of conventional thinking, which claims that stallions often kill foals who are not their own.*

I was surprised.

"What," I asked him, "are the mares getting up to when no one is looking?"

He answered with the backstory of High Tail, a seemingly nondescript little mare, a plain Jane with sagging back and poor coat, a mare we were watching at the foot of the Pryors, on the Wyoming side of the mountains. She was dubbed High Tail because the dock of her tail sat a bit too high on her croup. An aging dun with a thick, solid black stripe running down her back, High Tail had zebra-like black bars on her withers and zebra-like striping on the backs of her lower forelegs. Apart from these distinctive markings, High Tail looked like any mare standing in a farm field. If you didn't know her life story, you could easily mistake her for a child's riding pony or a retired plow horse. With her glory days clearly over, you probably wouldn't give her a second glance.

Yet Ransom's data showed that this mare had had a rich and varied life that involved a number of long-term male associates. She had been deeply bonded to at least one of them, her youthful first attachment. High Tail was by no means as physically powerful as stallions like Duke or Tecumseh—but what she *did* have was an abundance of persistence. She called her own shots.

Free-roaming horses tend to have minds of their own. Says Phyllis Preator, "They think different. That's all I can say. They just think different." If mustangs in general "think different" from domesticated horses, then High Tail tended to "think different" even from other mustangs.

Many of the Pryor Mountain horses choose to spend their summers in the flower-filled lush meadows thousands of feet above where we stood watching High Tail. Those high summer meadows, filled with aromatic lupines and other delicious treats, are like the proverbial land

* Stallions do sometimes kill foals from other sires, but no one knows how often this happens or why it occurs.

of milk and honey for horses who must weather unpredictable and harsh Wyoming winters.

Nevertheless, High Tail never went up there. She was born in 1989 down in more desertlike regions and that's where she chose to stay. That's one of the major differences between horse bands and grazing herd animals. Horses prefer a familiar home territory. They circle around from spot to spot, enjoying windy ridgelines in the summer and protective valleys in the winter, but they rarely migrate long distances.

Ransom first caught up with High Tail in 2003. He found the mare passing her days in the company of Sam, a stallion born in 1991. They made up a pair who, Ransom thought, probably encountered each other during the wanderings of their youth. An old myth claims that a stallion acquires mares, but if you watch closely enough, you'll see that mares sometimes work hard to get a stallion's attention. They can be every bit as assertive as stallions.

However the alliance between High Tail and Sam began, it worked. They stayed together for years. Eventually, other mares joined them and Sam found himself attached to a small mare-and-offspring group. Research shows that roughly half the time mares and stallions bond in this peaceful fashion. There's no need for a stallion to "conquer" the mare; she's often a more-than-willing partner.

Shortly after Ransom began following High Tail and Sam's band, he noticed a second, younger stallion hanging around a short distance away. Sam did not welcome this new stallion, dubbed Sitting Bull. The more Sitting Bull tried to become part of the group, the more Sam fought him off. Sam spent a good deal of energy trying to drive away the younger stallion, but to no avail.

Whenever Ransom saw High Tail's band during this period, Sitting Bull was usually there, hanging around on the outskirts, stalking the mares and dogging Sam, waiting for his chance. A stallion like this, called a satellite stallion, adopts a mating strategy of patience. He's always there, just on the outskirts, hoping to catch the eye of one of the mares: "To be a stalker pretty much sums up the idea," Ransom said. There are stories in the scientific literature of satellite stallions learning how to cooperate with the lead stallion and thus gradually gaining the ability, on a limited basis, to mate with some mares, but this was not the case with Sam and Sitting Bull. The two fought continuously. Still, Sitting Bull stayed near, biding his time.

His chance came in 2004. Horses who live at the base of the Pryor Mountains constantly face the challenge of finding freshwater. High Tail's band often descended the steep walls of the Big Horn Canyon Gorge, where they could drink their fill of river water. One day, they went down as a group. Records show that Sam did not allow Sitting Bull to come along. While the young stallion waited above, the rest of the horses stood on a small ledge and drank. Off in the distance heavy rains broke out. A flash flood inundated the gorge, cutting off the animals' escape route. For about two weeks, High Tail and her band, along with Sam, remained trapped without food. Conditions were so stressful that one mare died giving birth.

Realizing that the situation was dire, people intervened and helped them escape. The severely emaciated animals managed to climb up out of the gorge. Sam in particular had lost his muscular sleekness. Almost dead from starvation, he was easy pickings for the satellite stallion, who had hung around above the gorge. When the horses came up, Sitting Bull "just swooped right in and drove Sam off," Ransom said. Sam tried repeatedly to drive off his younger competitor, but he was no longer strong enough.

Most of the band accepted the young stallion. Not High Tail. The old dun mare chose Sam, the stallion with whom she had spent a good deal of her life. For High Tail, the bond with Sam was even stronger than the bond she had with the other mares in her group. At every opportunity she left her own band and headed off in search of her longtime mate. Each time she left, the young stallion chased her back, snaking his head and baring his teeth, threatening her with injury.

To avoid being bitten, she complied and returned to the band, but the next time Sitting Bull failed to pay attention, High Tail took off again. "We'd see her with Sitting Bull briefly," Ransom explained, "and then we'd see her back with Sam." This went on for many weeks until the younger stallion gave up chasing her. "From then on," Ransom said, "it was just Sam and High Tail. They got their weight back and at first

Sam tried to drive Sitting Bull off and get back with the other mares, but each time he tried, he failed."

High Tail stayed with Sam until he died in 2010. (Because of the stress of constant fighting with other males, stallions often live much shorter lives than mares.) Following Sam's death, researchers saw High Tail with a stallion they called Admiral. Eventually Admiral, too, fell from grace.

When we saw her that July afternoon, old High Tail was with only two other horses. One was a mare from her original band, an animal she'd known for years. The other was, ironically, that old usurper Sitting Bull. Rejected by High Tail in her younger years, he was *now* one of her boon companions. Primate field researchers long ago discovered the ebb and flow of alliances within primate troops, and finally we know that horses in the wild behave this way, too.

I asked Ransom if he thought there were any hard-and-fast rules about horse behavior in the wild.

"They rarely choose to be alone," he said. "That's one given."

Other than that, he couldn't think of much. Like humans, horses are blessed with an exceptional cognitive flexibility that allows them to adapt to a phenomenal variety of situations.

Traditionally, we've thought that horses only function via a kind of computerlike binary code of positive and negative reinforcement—the carrot or the stick. Now that science is showing us the subtleties of how horses naturally interact with each other, we can expand our own interactions with them, improve our ability to communicate with them, and enrich our partnership. This is exciting news, not just for horses, but for us, too. A relationship that has been traditionally seen as unidirectional—we command and they obey—can now become much more nuanced and sensitive.

I once acquired a dog who had been very finely trained. Whatever I asked that dog to do, he would do. This was fun at first, but ultimately, it was boring. There was nothing coming back from the dog, no emotional reward. After a year in my chaotic household, though, the dog had learned that our relationship was a two-way street. He was still obedient (to a degree), but he was much more of a developed individual. And he was a whole lot more fun to be with.

I suspect that I enjoyed Whisper for the same reason: He was, as I said, the most polite horse I ever had. But he was also very independent, very much his own man. He taught me more than I could ever have taught him. Whisper's course in why it's useful to think about the minds of animals has stayed with me for a lifetime and, quite probably, kept me alive in some pretty dicey situations when I found myself on the back of a horse in a place where I definitely should not have been.

But I never expected to get another life lesson from a horse out in the mountains of Wyoming. And yet, there was that old dun mare, High Tail, sending the message that a life well lived may have nothing to do with glamour, power, and drama, and everything to do with just being quietly persistent. Fate had not favored her. Nevertheless, she had lived long and prospered, had several long-term relationships and a number of offspring. Perhaps evolution sometimes favors the nondescript.

As Ransom and I talked, in the distance we could see the various ranges of the Rocky Mountains, mountains which, ever since they began regrowing 66 million years ago, have dominated the story of horse evolution. As we watched the old mare, the sky darkened over the peaks of those mountains.

For about the umpteenth day in a row, temperatures had stretched up toward 100 degrees, and over in Cody, the sun was relentlessly baking any living thing that dared emerge from under a rock or a roof or a leaf. Walking through the sagebrush from Ransom's truck to find the horses, we'd worn nothing but light T-shirts, and even those felt too heavy beneath the blazing sun and deep blue sky.

But now, over those ancient mountain peaks, midnight-black clouds rushed in. Up on the nine-thousand-foot-high pinnacles where other bands of horses were munching on summer flowers, we saw bolts of lightning strike the ground at intervals of only a few seconds. Ransom didn't appear bothered by this sudden change in the weather. I was, though. In New England, where I live, this sort of meteorological misbehavior is frowned upon. A sudden change from a bright blue sky to a sinister jet-black firmament seems to New Englanders to be rather apocalyptic. But out here, folks accept these climatic extremes as just another part of life.

Like Ransom, High Tail seemed not to notice the sudden atmospheric shift. Using her dexterous lips and well-worn front incisors, she just kept picking and choosing which of those sad-looking tufts of brush she would opt for next.

Surely, I thought, this is no place for a horse. I was thinking of the green pastures of New England and of the beautiful meadows higher up in the Pryors.

"Why is she here?" I asked.

"I don't know," Ransom answered. "There must be a reason. We just don't know what it is."

By this time the storm that had charged over the Rockies had brought swiftly plummeting temperatures. Raindrops the size of jelly beans began to fall. The drops changed to hail. We ran for the truck. High Tail just kept on eating. If the Vogelherd horse is the quintessential stallion, High Tail seemed to me to be the horse as survivor, capable and competent, steady and dependable, and able to adapt no matter what life might throw at her.

Indeed, wild horses are experts at adaptation, whether within the context of their own individual lives or over the generations, evolutionarily speaking. They don't care about sudden temperature changes, massive hailstorms, frigid weather, dry heat, or falling snow. Horses can live almost anywhere. It takes a lot to bring them down.

To think about this, all you have to do is consider the plentitude of free-ranging horse bands all around the world. No one knows exactly how many such horses roam our planet or even how many regions have free-ranging horse bands, because these horses often find their way into nooks and crannies where humans, even today, rarely tread. Bands of horses can thrive at ten thousand feet above sea level, or on isolated, hurricane-prone Atlantic islands. They can revel in fields of Kentucky bluegrass, or live in deserts, or make do with a diet of sea peas and beach grass.

When I started researching free-roaming horses, I was astonished at their numbers—in the millions. I was also surprised by the variety of ecosystems where the horses not only live, but thrive. There may be more than a million free-ranging horses in the Australian outback alone, living in conditions so unfathomably harsh that they make High Tail's grazing grounds seem like paradise. The Down Under horses, called brumbies, are tough—maybe even tougher than High Tail. There's an old horsemen's myth that horses do not do well in heat, but that's certainly not the case in the Australian outback, where temperatures can easily top 100 degrees for days at a time. This is certainly evolution at work.

Those horses too fragile to handle life in the outback die young, leaving no offspring. Only the most rugged, independent, and intelligent make it through this purgatory and survive to create the next generation. (Whisper, my street genius, would probably have made it. Gray, maybe not. Or maybe he would have bonded with a buddy as smart as Whisper and thus survived.) The Australian brumbies differ from domestic horses particularly when it comes to their feet. Over the century or so that they have lived in the outback (there were no horses in Australia until a stallion, four mares, a colt, and a filly arrived along with a shipload of English convicts in 1788), they have evolved exceptionally strong hoofs that can cope with constant walking over some very abrasive surfaces. One Australian researcher followed a band that traveled for two days to water where the horses drank their fill, then traveled back for two days to where grazing was available. The researcher saw a newcomer, an escaped domestic horse, join up with this band. He didn't last long. Unable to keep up the pace, he suffered greatly and died. Evolution sometimes happens this way. The horse probably had many great qualities—but he didn't have the *right* qualities to succeed in that particular environment. After Darwin published his theory of evolution in the middle of the nineteenth century, many people interpreted him as having said that only the strong survive. Darwin's thinking was much more nuanced. While he believed that the "struggle for existence" sometimes involved what he called "warfare" between species, he also understood that, like the poor domestic horse who couldn't survive in the world of the brumbies, animals not well suited to specific environments simply would not leave offspring. Consequently, a species, given enough time, would change.

It turns out that horses are quite talented at changing as the world around them changes. There are a few other places where horses have adapted to live in exceptionally dry areas. In the Namibian desert in southwest Africa (another place where modern horses did not live until they were brought there by humans), free-roaming horses have thrived in harsh conditions for almost a century. Researchers think they may descend from horses used by German soldiers when the region was a German colony.

All over the American West, free-ranging horses roam in small bands. They even seem to do well in areas around Death Valley, one of the hottest and driest places on Earth. You would think that a species that can live in Death Valley would have trouble living in swamps and wetlands, but it turns out that they don't. A little south of the Namibian desert, another population of horses lives in the Bot River delta of South Africa. And along the Atlantic coast of North America, from Cumberland Island National Seashore in Georgia north into Canada, horses live on numerous sea islands, including those of North Carolina's Rachel Carson Reserve and Currituck Banks Reserve, and on Assateague and Chincoteague Islands in Maryland and Virginia. Local lore claims that the ancestors of these horses swam to these islands after the many Spanish shipwrecks that occurred in the region during the 1500s. The equine geneticist Gus Cothran has shown that there is indeed a genetic connection between these island horses and those in Galicia studied by Lagos and Bárcena.

However the horses got to their islands, local people over the centuries let them stay. Such abandonments happen frequently. It's expensive to keep horses in barns and in paddocks and to feed them; those same horses, left to their own devices, can feed themselves quite adequately. In recent years, in areas of the American West that have experienced economic decline, domestic horses have been abandoned to fend for themselves—a tradition that's as old as horsemanship itself. In Europe, when the Soviet bloc broke up in 1989 and 1990, before leaving the collective farms where they'd been forced to live, Romanian farmers released their workhorses onto the Danube River Delta, to give the animals a fighting chance at survival. They did much more than just survive. Twenty-five years later, those horses, possibly the world's newest population of free-roaming horses, now number in the hundreds, and various local factions and international animal welfare groups are fighting with each other over how to manage them.*

The evolutionary resilience of horses is still manifest in our world today. Fifteen hundred miles west of the Danube River Delta are the famous white horses of France's Camargue wetlands, widely believed to be one of the world's oldest free-roaming populations. Some people say these coastal Mediterranean horses derive from Roman stock, but others say the horses have "always" been there, perhaps since the days when Egypt reigned. (Or maybe even longer.)

The Camargue delta is hot, humid, and filled with all kinds of disease-bearing insects. It's the kind of world where horses shouldn't be able to live. But they do. What's even more unexpected is the fact that the Camargue horses are white. Their coloring puts them at greater risk for diseases associated with sunlight, such as skin cancer. (This vulnerability is one reason why European royalty kept white horses: their coat color spoke of the wealth of their owners, who could afford to pamper their horses indoors and feed them, rather than putting them out to graze.)

Scientists who study evolution have long wondered why these horses are white. Wouldn't evolution have weeded out animals with delicate skin? A hypothesis that white horses are less visible to predators in open marshland than dark-coated horses turned out to be wrong, as did a hypothesis that white coat color kept the horses cooler by reflecting more sunlight.

The answer lay in the presence in the Camargue of immense numbers

* Some sources suggest that the horses from the collective farms may have joined a population of free-roaming horses present in the region for several hundred years.

of very large flies, insects that in great enough numbers can kill a horse. Carriers of infectious diseases, horseflies in great quantity may suck so much blood from a horse that the animal becomes ill. In large enough numbers, horseflies can literally annoy horses to death by preventing them from eating.

The Hungarian researcher Gábor Horváth and his colleagues have found that, in this situation, white horses have an advantage: flies attack white horses substantially less often than dark horses. Horseflies target their prey by following polarized light, in which the photons oscillate up and down in the same direction rather than every which way. Darker horse hairs polarize sunlight more efficiently than lighter horse hairs, so the signal the flies are looking for is stronger when the horses are darker colored. White horse hair polarizes almost no light. Ergo, horseflies bother white horses less.

So why aren't most free-ranging horses white? In most parts of the world, horses enjoy a number of options when it comes to avoiding horseflies. They can stand on ridgelines where the wind blows, discouraging the insects. They can graze in drier regions during horsefly season. (Dehydration kills the flies.) Or they can stand in cooler areas, also discouraging the flies. But in the Camargue, there are no options. The delta is horsefly paradise, with its perfect trifecta of conditions—the right amount of humidity, the right amount of heat, the lack of wind. Consequently, here the white coat color is an advantage rather than a disadvantage and is, therefore, selected for.

Like other scientists interested in understanding how systems change over time, Charles Darwin spent much of his life contemplating the adaptability of horses over millions of years, but it's only recently that science has shown us how evolution has helped horses in the modern era by giving them such flexible genomes. The same species that evolved to live on cold, dry Ice Age plains, like the Vogelherd horse, has also evolved to live on the shores of the Mediterranean in the hot and humid Camargue.

Or consider the case of the sea-island horses who live on Canada's

Sable Island, a small harborless sandbar of an island located far out in the North Atlantic, about a ninety-minute plane flight east from Halifax, Nova Scotia. This tiny island, shaped like a crescent moon, is about thirty miles long and very narrow. Buffeted constantly by violent North Atlantic storms, this island seems an unlikely home for free-roaming horses, yet as many as 450 graze here, surviving by eating beach grass and sea peas. This sounds like a meager diet, but the horses, abandoned there by a Boston entrepreneur before the American Revolution, have endured for more than 250 years.

Predator-free, their numbers rise and fall according to environmental conditions. No one feeds them or takes care of them. Since the 1960s, their numbers have not been culled. The Sable Island horses, in fact, may be the modern world's *only* genuinely free-roaming, entirely unmanaged horse population.

The horses are dependent on whatever gifts the sea provides, yet in recent years their numbers have increased. When I met him at the Vienna conference, the Canadian researcher Philip McLoughlin told me, surprisingly, that he suspects that the population explosion of horses may be due to an explosion in seal numbers. It seems that following a global prohibition against seal hunting, several hundred thousand seals now give birth on Sable Island yearly. This pupping, McLoughlin theorizes, with its accompanying deposition of an awful lot of fecal matter, has increased what he calls "sea-to-land nutrient transfer." In other words, all those seals leave behind a whole lot of nitrogen-rich manure. The manure feeds the plants. The plants feed the horses.

The only non-marine mammals on the island, the horses serve as a real-world laboratory of evolution. Over the centuries, they have become unique. Their pasterns are now so short that, from a distance, their lower legs look something like the legs of mountain goats. The pasterns of most horses are long and angled, allowing for plenty of spring in the horse's step, which in turn allows for greater speed and stamina when a horse gallops at high speeds over an open plain. Long pasterns evolved as a survival strategy. But longer pasterns also carry an important disadvantage: the pastern's fragile bones and vulnerable tendons can easily break or strain, laming the horse. Many a racehorse has ended his career because of this vulnerability. But on Sable Island, the horse does not have to run fast to escape predators. Instead, their enemy is deep sand and their worst "predators" are steep, treacherous sand dunes, some almost a thousand feet high, which the horses must climb in order to eat. These dunes provide some pretty dangerous footing for horses. On Sable Island a horse is much more likely to injure a leg while descending these steep dunes than by running along the island's beaches. Still, a hungry horse must ascend and descend these obstacles.

Consequently, evolution has made a clear choice, just as in the Camargue region. Sable Island horses have shorter, less vulnerable pasterns, giving them that goatlike look. Over 250 years, natural selection has opted for shorter pasterns, improving the horses' ability to graze, thus improving the horses' ability to live longer and produce more offspring. We often think of evolution as complicated, but in this case, the process is pretty easy to grasp.

The Sable Island horses are also behaviorally unusual. Worldwide, many horse bands share their home ranges with other bands. Although the bands don't travel together, their ranges often overlap, leading to the squabbling Ransom and I watched in the Pryor Mountains. While the horses constantly argue and fuss with each other, normally they don't drive the other bands away from specific grazing areas. They are not territorial.

On Sable Island, however, they are very territorial. Island resources are limited. Under these conditions, the horses stake out territories which they defend from other bands. Rather than sharing grazing opportunities, McLoughlin has found, the horses divide the island into three distinct territories: the western end, with its rich grazing and year-round freshwater pond; the middle area, with poorer grazing and ponds that are often full; and the eastern end, with very poor grazing and almost no surface water. The horses on the eastern end must use their hoofs to dig holes into the sand to reach freshwater. The horses on the island's western end do not allow the horses from the eastern end into their territory. In effect, the horse bands have formed hierarchies, giving new meaning to the concept of a stratified population of upper, middle, and lower classes.

Elsewhere in the world, other populations of horses have evolved other unusual environment-dependent survival strategies. Siberian horses living in the Yana River region, near the Arctic Circle, sometimes called Yakut horses, are said to enter a kind of quiescence in the winter. During the summer, the horses graze constantly on steppe grasses, storing up calories in a subcutaneous fat layer. In the winter, the horses survive by slowly burning this fat, by breathing only half as much as they do in the summertime, it is said, and by standing quietly without exerting themselves whenever possible. Villagers think of this as a kind of semi-hibernation.

Innovation is a never-ending process. Around the world, wherever they live, given enough time, horse bands tend to become distinct entities, well-honed to the world they inhabit. Watching them is an invitation to fall in love with their backstories—not just of individual animals, but also of their long-term evolution.

The domestic horse is a generalist, bred by us for millennia according to our own wishes in order to meet our own needs. One reason why this has worked so well for us is because of the horse's evolutionary malleability. We've been blessed by this flexibility. We've been able to breed huge Shire horses, capable of carrying knights into battle; quarter horses with heavy haunches who can spin on a dime; light-boned trotting horses who can pull carriages; and lithe ponies able to jump high fences despite their short legs.

Where did all these qualities come from? We do know that *Equus*, our modern horse and "the best running animal on the planet," according to the paleontologist Darrin Pagnac, appeared at least 4 million years ago. But he's a marvel descended from 56-million-year-old ancestors who arrived on our planet in a far different form. Since that time, he has been shaped and reshaped by tens of millions of years of global heat spikes, fluctuating ice ages, tectonic upheavals, volcanic mega-explosions, and many other planetary forces until, today, he has mastered the art of adjusting. Extremely intelligent, he can fend for himself in the most challenging environments or be coddled in our barns, pastures, and twenty-first-century cities. He can, like Whisper, learn new skills and cope with problems even when the barn manager, like me, leaves much to be desired.

All horses large and small stem from only a few Pleistocene survivors.
(arjecahn, Catch Me If You Can, February 4, 2006, via Flickr,
Creative Commons Attribution)

"Horses today," Ransom once told me, "occupy an anthropogenic niche. They'll live wherever we let them live. If you give them enough space, they'll figure it out."

One long-ago afternoon in California's Death Valley, I watched two free-ranging horses standing very still, trying to weather the midday sun. Death Valley is named for the potential effect of its blistering heat, and when temperatures are at their worst, you want to put a damp cloth over your face when you breathe to keep from searing your lungs. The horses stood stock-still, surrounded by hot sand. Under the merciless sun, I understood their behavior. Like them, I moved as little as possible.

It was too dry for horseflies, but some insect was bothering the animals. They stood head to rump, a bonded pair, swishing their tails back and forth to help each other out. I'd seen this behavior a thousand times before in my life and never given it more than a passing thought until then.

But something struck me that day. Despite the miserable heat, I wasn't alone. Other horse watchers were there. It was then that I realized what an inestimable gift horses are to people in the modern world and how bereft we would be without them. Even today, long after that ivory carver created the Vogelherd masterpiece, we take so much pleasure in watching wild horses that, all around the world, we let them roam free, giving them their own unique status—part wildlife, part domestic livestock, part companion animal, part guide into the mysterious world of nature. We need horses in our lives.

In one barn where I rode, a retired couple brought their rescued horse with them for a few weeks of vacation. I thought this so charming that I asked them about it. It turned out they never went anywhere without their horse. He was so deeply bonded to them that he became overwrought if they weren't there. I suspected that the partnership was a two-way street and that the process of caring for the animal, which in their case involved an elaborate two-hour daily grooming ritual (he had his nostrils and ears sponged out daily and in the course of this enjoyed many carrots), was equally soothing to them.

Every morning the husband and wife showed up to brush the horse, clean his teeth, and wash his eyes. They talked to him and handed him treats. Apart from the Lipizzans I visited at Vienna's Spanish Riding School, this gelding was the cleanest horse I'd ever seen. He was rarely ridden, but when he was, the wife rode the local trails on the horse's back and the husband accompanied them on his motorcycle. The parade was a wonder to behold.

Horses evoke something ineffable in the human psyche, something at once both exciting and calming. Just looking at a painting of horses, on a museum wall or on a cave wall, can be heart-stopping. Their presence in our lives makes the world so much more grand, even if we only see them from a distance. When the U.S. National Park Service wanted to remove some horses from protected riverbanks in the impoverished Missouri Ozarks, the mountain people objected. The horses themselves were nothing unusual, having apparently found their way onto national parkland after being abandoned by farmers during the worst days of the 1930s Depression. These horses were just like the horses that locals had in their own barns and pastures. But still, many in the region wanted the horses left alone. They found comfort in their presence.

"As long as the wild horses continue to roam," said one man, "then maybe there's hope for us as well."

Perhaps this is the significance of the horses created by the Ice Age artists: They represented hope. And just simple companionship. What are the roots of such a partnership?

2

IN THE LAND OF BUTCH CASSIDY

> If you want to sense the evolution of the modern horse, you can grasp a horse's fetlock and still feel the remnants of toenails that his ancestors had.
>
> —RICHARD TEDFORD, American Museum of Natural History

One day while I was still in Wyoming, Phyllis Preator drove me in her heavy-duty, fire-engine-red ranch truck, complete with American flag and rattlesnake pistol, up to the top of Polecat Bench. It was late afternoon, and on our way up we passed a line of oil field roustabouts and roughnecks getting off work and heading in the opposite direction down into town.

There's a lot of oil in this region. Long before the modern Rocky Mountains started growing about 66 millions years ago, seas covered what is now Wyoming, off and on, ebbing and flowing with continental drift, tectonic collision, and changes in global and local climate conditions. The prevalence of these shallow and well-lit seas in which large and small marine organisms flourished helps to explain why early wildcatters found in these High Country deserts layers of "shale so black it all but smelled of low tide," in the words of the inimitable John McPhee. Because those ancient seas ebbed and flowed, then ebbed and flowed again over tens of millions of years, layers of earth are now interspersed with pockets of oil—dead, decayed, buried, and cooked-by-the-heat-of-the-planet sea life compressed into a fluid form of carbon that's easy for us to transport and to burn.

The sea life that transformed into that fuel varied from place to place, so the oil from each region has its own thumbprint of sorts. The oil pumped out of Polecat Bench stinks. Distinctively. It's seriously sulfurous. Hence the name Polecat, which is Wyoming for "skunk." Preator's friend Nettie Kelley, who had tagged along for our day of wild-horse watching, explained that when the oilmen working there come home, "the first thing their wives do is wash their clothes. They can't stand the smell." Breathing in the dust, I had a hard time imagining the place covered by early oceans or the swamps that carpeted it after the dinosaurs who lurked here in great abundance became extinct (except for the birds).

Preator had her ever-present baseball cap smashed down over her medium-length blond hair, which stuck out the sides like feathers. She wore a shirt covered with miniature galloping horses and with lots of words like "spirit" and "stallion" and "free" and "beauty" written in cursive. Busy with family and bringing in the hay and keeping her own horses fit, she hadn't been working on her *Sage Brush Annie* blog lately (the nom de plume derives from her admiration for the original Wild Horse Annie who advocated for mustang protection in the mid-twentieth century), but the horses roaming the range still pluck at her heartstrings. Now in her mid-sixties, she remembers that when she was just five years old her father brought her first horse into the kitchen for her to meet. Since then, she's never been without a horse. Or two. Or six or seven. (Preator is not alone. The owner of a bed-and-breakfast where I stayed told me he had adopted eleven range horses: "I don't have a dozen, because that would be excessive.")

We had been watching horses on McCulloch Peaks and now we were replaying what we'd seen. The horses had been all bunched up on a ridgeline, using the cooling breezes to get rid of the flies. Thinking about how much time barn managers spend trying to ensure that horses pastured together don't hurt each other, I wondered how these guys could get along in such tight circumstances without human oversight.

"They're always talking to each other," Preator said, reminding me of our "think different" conversation. "They'll draw the proverbial line in the sand, then stand on one side and stand on the other side and treat it like it's a backyard fence. They'll line up and talk to each other for hours like that. They're always working things out." There's a lot of snorting and stamping of feet, but injuries are rare. Even the stallions rarely resort to extreme violence, although screams and biting are a common aspect of the negotiations.

An eminently practical and pragmatic Westerner—she once butchered a bison by herself in the backcountry with only a knife made from obsidian, just to make a point—Preator is also a definite believer in the competence and integrity of free-roaming horse bands. And she has a soft spot for the Peaks horses, recent arrivals descended mostly from saddle and draft stock brought in by early ranchers. She has even meticulously researched and written their history by interviewing a number of old-time ranchers in their later years. The result is her book *Facts and Legends: Behind the McCulloch Peaks Mustangs.*

The Pryor Mountain mustangs with their Spanish coat colors may get a lot of great press, but the Peaks mustangs have genetics you won't find in the Pryor Mountain animals—strains of imported English Shire horses and of French Percherons and some Morgan horses and even, maybe, a few Thoroughbreds. It's thought by local cowboys to be a good thing to have a Peaks horse. Hardened by life in this rugged area, Peaks horses are said to be high-quality, versatile mounts. Cowboys used to earn pocket money by rounding up the best of them, putting a bit of training on them, and then selling them on.

By the time we got up on top of Polecat Bench it was scorching hot, about a hunnert degrees ("hunnert" being a meteorological term commonly used throughout the American West), just like everywhere else along the Front Range that summer. The wind blew almost seventy miles an hour, not in gusts but steadily, like the flow of a mighty river.

Preator and Kelley assured me that this was typical Wyoming: "Windier than the state of Wyoming is what we call people who talk too much out here," Preator said.

I got out of the truck and the wind ripped the door handle from my clenched fingers. I'd hoped to sit and enjoy the view, but quiet contemplation was out of the question.

Getting sandblasted by the wind wasn't what I'd had in mind. I must have looked annoyed.

"Around here we call this easy chair weather," Kelley said. "Just a little bit harder and we could lay back in it and take a nap."

Preator and Kelley and I had spent most of our day watching mustangs on McCulloch Peaks, but I had also wanted to drive up here to Polecat Bench, to see a certain layer of spectacularly bright red dirt that contained the world's earliest known horse fossils. The bench is famous among local kids as a great place to party and famous among paleontologists as the world's best place to find mammal fossils from a certain age—about 10 million years after most dinosaurs died out at the end of the Cretaceous period.

At that time, roughly 56 million years ago, the world's first horses—the dawn horses—show up in the rock record. Fossils of these horses have a very special status in paleontology: soon after the dawn horses arose, they spread prolifically and ubiquitously, so that their fossils are found in great numbers in many places in North America and, in slightly different versions, in Asia and all over Europe.

Dawn horses are thus a paleontological page marker, an "index fossil." Their presence at a dig site is like a chapter heading indicating that the researcher is reading in a special layer of pages in Earth's history, in a remarkable epoch of time called the Eocene—the dawn time, the beginning of *us*, the explosion of life as we know it today, the beginning of the age when modern mammal families came into their own and spread worldwide. If it seems coincidental that the earliest known horse is found at the beginning of the Eocene—it isn't. Paleontologists have marked the beginning of this epoch of time in accordance with the date of appearance of the earliest known horse. That's how important horses were, and still are, to the science of paleontology.

Once horses turned up on Polecat Bench, they were suddenly everywhere. There are so many of these dawn horse fossils in so many Northern Hemisphere sites that we are certain that, even during these early days of horse evolution, horses were already a fabulous idea, a true success story. Small as they were, often well under two feet high at the withers, they made up in total body mass what they lacked in individual stature. In some places, they seem to have sprouted forth like weeds. In certain parts of the world, if you're looking in the right geologic time layer, it's difficult *not* to find them. Especially if you have Kelley's easy-chair wind doing most of the uncovering for you.

Paleontologically speaking, Polecat Bench is special. In a sense, 56 million years ago the site was like a little nest. To the west, the modern Rocky Mountains, poised to play a key role in horse evolution tens of millions of years later, had already grown enough to at least slightly disrupt the winds that blew from west to east across North America. To the east the Bighorn Mountains were beginning to appear. Other mountain ranges surrounded the area, as the North American tectonic plate pushed relentlessly against the tectonic plates of the Pacific Rim.

In the midst of these mountains sat the region now called Polecat Bench, well-watered by runoff from the surrounding mountains. The bench today is like a tabletop that you have to drive up onto, but in those days, before rivers eroded much of the surrounding land, it was a riverine region with a variety of habitats. It seems to have been a very nurturing location. Along with the earliest known horse fossil, paleontologists have found here in the same time period the earliest known fossil of a euprimate—a true primate.*

So where I was standing on Polecat Bench was the spot where the deep-time foundation for the partnership between Whisper and me took a major step forward. Stretching my mind between this past and my own present was dizzying, like trying to imagine infinity when I was six years old. Right where I was standing, a dawn horse might have nibbled on ancient grapes.

This joint appearance of horses and primates, together in the same locale and in the same deep-time frame, is not coincidental. In those early Eocene days, we both enjoyed the same damp, hot, junglelike environment, which isn't surprising given that we shared a common ancestor in the probably not-too-distant past. There's plenty of evidence for this shared ancestor, but the easiest way to grasp this truth is to consider that we share a common skeleton, albeit one stretched in different

* It's important to differentiate between a "true primate" and earlier primate-like creatures, which are found in more ancient rock layers and are often written about in the popular press as "primates."

ways and to different purposes. Today we look dissimilar, but there was a time when we could have been mistaken for siblings.

We can see vestiges of that kinship in our own skeletons. For example, the ancient horse, the modern horse, the early primate, and the modern human all have a patella. In humans, the patella is also called the kneecap. You can feel it as a rounded bone on the front of your knee. It's held in place by tendons, and woe betide anyone who injures those tendons. The kneecap will no longer be able to do its job. In horses, this bone is called the stifle. It can also easily be felt, and woe betide any horse who damages the tendons that hold it in place. To find it, run your hand up the front of a horse's rear leg, all the way up his leg until you nearly reach his belly. There you'll find the same bone you have in your knee. But in the horse, the bone feels not rounded—but pointed. I remember as a child feeling the stifle with my fingers and wondering why it was so strangely shaped. An early version—the beta version if you will—of the modern horse's stifle can also be found in the Polecat Bench dawn horses.

Another bone modern horses and humans have in common is the calcaneus. In the human, this is the heel bone. In the horse, it's the bone at the point of the hock. Run your hand up the back of a horse's hind leg. About halfway up you'll come to a joint between the lower leg bone and the upper leg bone. When I was a child, I thought of this as a kind of backward-facing knee, but I was wrong. It's not the equivalent of our knee, but of a bone that for us is in the foot.

As early as 56 million years ago, these differences had begun to take shape. The primate had a heel bone that pretty much looked like our modern heel bone and that most of us today would easily recognize as such. The dawn horses had hocks that had already become somewhat similar to the hocks of modern horses. If you knew where the hock of a modern horse was located, you would easily be able to pick out the calcaneus in a dawn horse.

This is very cool stuff. Horses and humans share tarsals and metatarsals, fibulas and tibias, and indeed, almost all the same bones—evidence of our biological kinship. Sometimes I imagine setting a skeleton of the first true primate and a skeleton of the early dawn horse side by side—and then running the clock backward, figuratively speaking. As we moved backward in time, we would see both skeletons become more and more primitive. The differences in the two skeletons would slowly disappear until, ultimately, meeting at the bottom of an upside-down triangle, they would be the same. We would be the same animal. Scientists call this the "stem" animal from which both horses and humans evolved. In other words, evolution—the great unfolding of life on our planet—is the foundation of the horse-human partnership, the reason why we can learn to understand each other so well.

It's not clear when this stem animal, our common ancestor, lived. Some researchers suggest that horses and humans parted ways and started down separate evolutionary pathways only shortly before they turned up at Polecat Bench, probably just after most dinosaurs became extinct. Others suggest that the parting of ways may have taken place earlier, about 100 million years ago, during a period romantically dubbed the Cretaceous Terrestrial Revolution. I first read about this revolution when I was just a child, in an essay entitled "How Flowers Changed the World," in *The Immense Journey* by Loren Eiseley.

This revolution, which was clearly under way 100 million years ago but which may have begun tens of millions of years earlier, was one of the most important upheavals ever to occur in the history of our planet—bar none. It was certainly much more important than the fall of the asteroid, for example. Before flowers, Eiseley wrote, "wherever one might have looked, from the poles to the equator, one would have seen only the cold dark monotonous green of a world whose plant life possessed no other color." After the revolution, flowers were everywhere, color covered the planet, and the world would never be the same.

In fact, without flower power, the Age of Mammals itself might never have occurred. We mammals might have stayed small, inconsequential, and rather boring, just as we were for most of our 200-million-year history. Let's face it: for much of our time on Earth, we just weren't that interesting. For one thing, we mammals stayed hidden—the best defense we could come up with, given the mass of dinosaur flesh stomping around on the green Earth during that time. We probably ate mostly insects and grubs and worms, and we probably only came out at night.

We didn't eat fruits or grains, because there weren't any. We did not enjoy a glamorous lifestyle.

Most likely, we just tried to stay out of the way. We probably defended ourselves by becoming what the paleontologist Christine Janis once called "artful dodgers," animals who scurried around in the underbrush. While the dinosaurs reigned, mammals were little more than also-rans, tiny little ratlike creatures with runty snouts and four legs and five digits at the end of each leg. Sure, a few of us, like a wolverine-size *Repenomamus*, may have made it a habit to eat small dinosaurs (one fossil of this animal has been found with a young dinosaur in its stomach), but for the most part we were all potential and not much self-actualization. Our self-esteem was probably pretty low, and we probably would have greatly benefited from the help of a New Age guru—had any been around.

Then the non-avian dinosaurs died. So did many of our mammal cousins. Along with the dinosaurs, an estimated two-thirds of the then extant thirty-five mammal families became extinct. These mammalian extinctions, though, were not spread across the globe evenly. (Extinctions rarely are.) Rather than being global, they were somewhat glob-like. A bunch here. A bunch there. In North America's northern interior, including the area around Polecat Bench, extinctions were *highly* glob-like: as many as nine-tenths of mammal families may have disappeared.

But, for better or for worse, one creature's misfortune is another's opportunity. A huge hole appeared in the planet's tightly knit web of life. The end of so many life forms opened the way for the dance of communication between horses and humans to take its first baby step forward. Correlated with the 66-million-year-old disappearance of most dinosaurs was the fall of a large asteroid into what we now call the Gulf of Mexico. So without the asteroid impact that cleared the stage for a great evolutionary leap, horses and humans might never have appeared. Was the asteroid's fall the cause of the extinction? Or was the timing merely coincidental? Were there larger forces at work, like tectonic movements and shifting ocean currents?

Most likely, paleontologists suggest, the truth behind the extinction involves many factors. When the asteroid fell, the world was already changing. The great supercontinent of Pangaea had broken up and North and South America were slowly migrating west, creating an ever-widening Atlantic Ocean—an ocean that would become a major player in the appearance of humans and in the evolution of horses and in the flight paths of birds and in the pulsations of ice and rain and drought for the coming tens of millions of years.

These long-term events, the results of our always-convulsive, seething-with-energy planet, were probably more influential in the appearance of horses and humans than the onetime crash of a mere mega-asteroid. So although paleontologists do not debate that the impact occurred, its role remains a perpetual, often contentious source of disagreement. A 2010 paper in *Science* authored by forty-one researchers, mostly non-paleontologists, asserted that the asteroid was the extinction's "sole" cause. This in turn caused a great deal of snorting when paleontologists met for lunch breaks. Speaking for many others, the Yale University paleontologist Chris Norris called the emphasis on disaster as a major evolutionary force "asteroid porn." Popular descriptions of the effects of the asteroid impact, he explained, "have a feverish quality to them that verges on the unsavory." His point is well-taken: the worldwide climate had been changing for 10 million years before the asteroid fell. The dinosaurs were no more enjoying a steady-state world before the asteroid impact than we are today.

"Don't get me wrong," David Archibald, an expert in pre- and post-impact mammals, told me. "It wasn't just another bad day for the planet. It was *the* bad day." But, he added, trouble had been brewing in paradise long before then.

In any case, after the asteroid fell, a prehistoric landgrab began. A whole lot of unoccupied space was suddenly there for the taking. All you had to do was figure out how to fit in. We mammals were really good at that task. Those of us who remained alive quickly morphed into all kinds of new shapes and sizes. It must have been somewhat like the early days of the Internet, when the future was wide open and anything could—and often did—happen.

We don't know a lot about how mammals revved up to meet these new conditions, but we do know that during these 10 million years, plant life changed dramatically. The evergreen forests that covered much of Earth died back and deciduous, broad-leafed trees began very slowly to spread, according to ecologist Benjamin Blonder. The expansion of these plants, with their delicious and highly browsable leaves provided an abundance of easily digestible food.

This, in turn, encouraged mammalian innovation. A new type of mammal arose—a mammal that could feast not on insects but on flowers, fruits, shrubs, and even, to a limited extent, on the very few grasses that were able to take root. Lots of these mammal experiments would turn out to be failed ventures, evolutionary dead ends that evolved and died out in a paleontological blink of an eye.

But a few succeeded spectacularly. Horses and primates ranked high among the triumphant. (We probably shouldn't feel too proud of our achievement, though. In the words of Christine Janis, we were "victors by default.")

Preator and Kelley and I walked around Polecat Bench, wishing we could find some of those cool ancient fossils, but before long I gave up. Philip Gingerich, a world-renowned paleontologist and a fossil-horse expert, once told me that when he first came to the bench as a young paleontologist in the 1970s, the fossils were just lying there, right on the surface, but those easy finds are long gone. Preator hunted for tepee rings, circles of stones outlining the circumferences of tents pitched by pre-Europeans who loved to camp up here and take advantage of the great view, just as modern local folk do on summer evenings.

Kelley looked for landmarks up in the distant mountains—the natural signs her own ancestors used a century earlier to orient themselves in place and time when they first arrived in the wilderness to run sheep.

"There's the Horse's Head," she said, pointing to a large, easily seen snowpack up in the Ishawooa drainage of the Absaroka mountain range. "The reins and the nose are melting. That means the high water's done and you can take your livestock up into the mountains and cross the rivers safely."

I saw that she was right. Just below one of the tallest mountain peaks there was still, even in early July, a large amount of snow tucked into a ravine, and that snow did indeed look quite like a horse's head.

"I guess," I said, "horses just belong here."

It's hard to imagine this Wyoming land without horses. One way or another, people in the region have for centuries timed their own yearly rhythms to the rhythms of horses. Without horses, life would have been very challenging. When the Spanish brought domesticated horses to this region, Native Americans immediately recognized their value and became some of the world's best horsemen. And when Kelley's great-grandfather, a Mormon, came here at the end of the nineteenth century, horses meant the difference between surviving and dying. The outlaw Butch Cassidy, who lived in the area, was sent to jail for the first time as a horse thief. Given the life-or-death importance of horses, he was lucky they didn't hang him.

Even when Preator's father brought her to Wyoming as a child in the 1950s, horses were still essential. Sure, there were cars by then, but cars needed roads, and there just weren't a lot of roads in the Wyoming wilderness. For a while, her dad trained horses for a living. Then he became a wildlife ranger and spent his life in the saddle patrolling the Wyoming backcountry, where cars could not—and still cannot—go. Preator often rode along at his side. She and her pony covered a lot of ground. I suspect that nearly every inch of territory in these mountains has been visited, trod upon, sat upon, or at least looked upon by Preator at one time or another. To her, there's something quite satisfying in knowing that horses have lived for 56 million years on the land where Peaks and Pryor horses live today.

Of course, it's important to keep in mind that when the dawn horses lived here, Wyoming looked quite different. It was wet—so wet that it was covered with tropical foliage. There were no cold-weather trees. Instead of dust-filled air and dry, hard ground covered with sagebrush, there was a lot of mud and swampland.

It was hot. For a brief period, it was *very* hot, much hotter than when I visited. In fact, it was as though there was a sudden explosion of heat, as remarkable in its own way as the fall of the asteroid had been

10 million years earlier. Curiously, this explosion of heat also marks the appearance of Polecat Bench's horses and primates. This was a time when temperatures in some places shot up by 6 or 8 degrees Celsius in a very short time period, lingered at those heights, then, almost as suddenly, dropped back down. The cause of this heat spike remains elusive, but it may have been due to large bursts of methane that bubbled up from the deep ocean.

On temperature charts that track the rise and fall of heat throughout our planet's history, the heat spike looks to me like the outline of the Eiffel Tower. The anomaly is officially called the Paleocene-Eocene Thermal Maximum, PETM for short, but I prefer to think of it as the Eiffel Tower of Heat, with its sharp lines of ascent and descent that mimic so closely the graceful lines of the Parisian landmark. It's a weird event.

And it's doubly weird that both horses and primates may owe their existence, in part, to *its* existence: the spike marks the beginning of the Eocene, when not just horses and primates, but most modern mammal groups finally came into their own. Many of our major mammal groups trace their first appearances to this puzzling heat spike. It's as though the whole world had become a giant petri dish brought to a boil by a colossal Bunsen burner. And voilà! A world that was drowning in post-asteroid misery suddenly experienced a global spring.

Opinions vary as to whether horses and primates originated on Polecat Bench or traveled there from somewhere else. Some paleontologists believe that proto-horses migrated over from Asia at the beginning of the very hot time. One very early fossil of an animal that might have been an ancestor of the dawn horses has turned up in China. Others say horses originated in Europe and traveled west. Philip Gingerich suggests that horses may well have originated right there, right in pre-Wyoming Wyoming, just east of the Rockies, right in the same region where Kelley and Preator and I were standing.

There's even more disagreement over the origin of primates. Some researchers suggest we originated in North America, while others say Asia, and still others, Europe. In what I think of as the Huck Finn theory, another group proposes primates evolved in Africa, then rode the continental plate of India as it separated from Africa and traveled north to collide with Asia, where they dismounted and spread across the Northern Hemisphere while it was still a pretty warm place to be.

Theories abound, but current evidence shows that the earliest dawn horses and the earliest true primates made their grand entrance onto life's stage as a duo, one species living in the trees, the other browsing below. And if we weren't yet a matched pair on Polecat Bench, if we hadn't yet partnered up, we certainly lived as close companions.

They were strange little things, these dawn horses. They certainly don't look like animals fated to run the Kentucky Derby 56 million years later. But, as I said earlier, if you know what you're looking for, you can see in these early fossils some of the basic characteristics that say "horse" to us today.

I, of course, did *not* know what I was looking for. Since childhood, I had seen fossils of dawn horses in museums and never really understood why scientists thought of these strange little beings as "horses." What was it in these skeletons that said "horse"? I knew a little about some of the bones we share with horses, like the calcaneus and the patella. But I wanted to know specifics. When I looked at a fossil of a dawn horse in a museum, it looked to my uneducated eyes rather like a dog.

Why *isn't* it a dog?

I called up Phil Gingerich. He told me to check out the astragalus bone, present in dawn horses as well as in modern horses. The horse's astragalus bone is unique. It's the horse's equivalent to the talus bone in the center of the human ankle, which lets us rotate our foot in a circle. But the horse astragalus is shaped differently. Our talus, the second-largest bone in the human foot, allows us to easily change the angle of our feet in relation to our legs. It is the reason why we can use our feet to climb trees.

Remarkably, by 56 million years ago, the horse's astragalus bone had already become distinctively shaped. It's in front of the hock bone, or the calcaneus, and by the time the earliest known horse appeared, the horse astragalus was already shaped differently from the primate astragalus, which would become our talus. Even at this early date, the horse astragalus was deeply grooved and committed to limiting the movement of the horse's hock to forward and backward instead of in a circle. Thus the horse, from his earliest days, could not climb trees.

This was not necessarily a handicap for the horse. Instead of being able to climb trees to escape predators, he opted for speed as an escape strategy. By becoming deeply grooved, the astragalus bone limited the movement of the leg below the hock to only one plane, giving the horse a slight advantage—the ability to move forward faster than other animals. Thus did speed become, from the horse's earliest days, his primary defense.

Of course, speed is relative. You have to think of the horse's speed in the context of the time. Dawn horses couldn't gallop, but at least they could hop, sort of, when most predators could not. "To get away from trouble," the paleontologist Mike Voorhies once told me, "a horse wants to run." He compared that strategy to the strategy of a bison, who, with his huge head and curved horns, wants to fight: "I think it's quite important that, from the beginning, there's never been a time when horses had any cranial ornaments." So, because of the Polecat Bench fossils, we know that from his very first days, the horse was better equipped to choose flight over fight when danger appeared.

Horses weren't the only early Eocene mammals on their way to becoming cursorial (committed to running), but something else important set horses apart: on their hind feet, they were odd-toed. It was very *odd* to be odd-toed. Most ungulates were (and still are) even-toed. As early as the Eocene, plant-browsing cursorial runners had divided into two groups: the odd-toed perissodactyls (horses and their close kin) and the even-toed (cloven-hoofed) artiodactyls. This is how scientists talk about them, but it's not the number of toes per se that's important. Rather, in horses, it's how the toes bear the animal's weight.

Even at this early date, even on Polecat Bench, the middle toe of the horse's hind foot carried more than its fair share of the animal's weight. You can see this if you look closely enough: the middle toe of the three on the hind foot (the toe that would eventually become the hoof of the modern horse) was just a little bit bigger than the two outside toes. The even-toed artiodactyls had already taken a pathway that would prohibit them, forever, from evolving a foot with only one toe. The dawn horses kept that option open.

This large middle toe was key to the horse's survival. Over time,

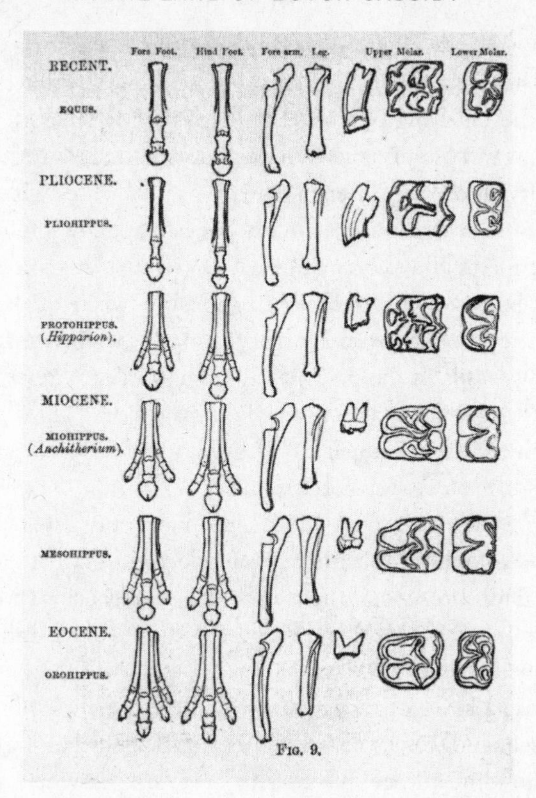

Diagram by the paleontologist O. C. Marsh of the evolution of horse feet and teeth over time

horses placed more and more weight on this middle toe, until the other ones became useless and disappeared. Ultimately, the result of this 56-million-year-old evolutionary decision would allow Whisper to throw his weight back over his hindquarters and take a graceful leap over my pasture fence in search of water. Quarter horses would be able to spin around after a cow. Dressage horses would be able to execute a *levade* or *courbette.* When the Vogelherd horse coiled his hindquarters and raised his back and neck, or when the Pryor Mountain stallions confronted each other, they owed their powerful abilities to this early evolutionary choice, traceable to the Eiffel Tower of Heat, to the early Eocene, and perhaps to the very spot on Polecat Bench where Preator and Kelley and I stood.

I know this now, but still, when I see the dawn horses, my heart aches for them just as it did when I was a child. They don't look comfortable. They had high arched backs that remind me of my border collie, who can twist up and down and to the right and the left with equal agility. The Polecat Bench dawn horses lack the stable backbone and high withers and straight cannon bones that allow a modern horse like Duke to stand so commandingly. I couldn't imagine what gaits these early horses would have had. Long before I visited Polecat Bench, I'd asked the paleontologist Margery Coombs, who was showing me dawn horse fossils in the paleontology museum of Amherst College, what she thought their fastest gait might have been.

She replied: "The scamper."

That didn't sound very noble to me.

It wasn't much by racehorse standards, but it was just about the best the age had to offer. It must have been good enough, because in these early days dawn horses and their other odd-toed cousins (precursors to modern tapirs and rhinos) were very common—much more so than the even-toed artiodactyls.

There are a lot of other ways in which these earliest horses do not seem horselike. Their necks were longish, but not long. They would have had to almost kneel in order to graze on grass, had there been much. Their necks were set at a low angle so that, had they yearned to look as proud as the Vogelherd horse, they wouldn't have been able to.

Their heads, too, were sort of horselike—but not very. The face was somewhat elongated, but still short by modern standards. The eyes sat toward the middle of the skull rather than closer to the ears. Modern horses enjoy nearly 360-degree vision, so that they can see what's behind them as well as what's ahead of them. (This is why carriage horses often wear blinkers; the glistening "predator" carriage following so closely on their heels terrifies them.) Dawn horses had vision that was much more limited.

Dawn horses also had padded feet, like cats and dogs. But at the end of each toe was a delicate little proto-hoof, a fingernail-like structure that protectively wrapped around each toe pad. This proto-hoof was too thin to bear the weight of the animal. Instead, the padded toes splayed out so that the pads rather than the hoofs bore the animal's weight. Just as the feet of modern moose splay out so the huge, heavy animals can navigate swamps, the earliest horse toes provided wide footing, so the dawn horses wouldn't get sucked into the swamps where they lived.

This is an interesting point: horses originally evolved to live not on hard, dry land, as in modern Wyoming, but in muddy, wet regions, maybe something like the jungle regions that tapirs, close cousins of horses, still inhabit today. Perhaps this muck-oriented deep-time history helps explain why horses can live today in regions such as the Camargue or on Atlantic coastal islands.

Most of us wouldn't easily recognize dawn horses as being ancestors of our modern horses, but what about the first euprimates, the little creatures who kept the horses company? We laypeople may not easily see "horse" in the dawn horse fossils, but it's easy to recognize Polecat Bench primates as our ancient kind. We don't need to have the similarities between their skeletons and our own pointed out to us in museums. The relationship is pretty obvious. In *The Beginning of the Age of Mammals*, the paleontologist Ken Rose compared these early primates, already adapted for leaping and grasping, to today's bush babies. The euprimates weighed only a few ounces and had teeth smaller than grains of rice. Their extravagant tails likely provided balance as they moved from tree branch to tree branch. They already had proportionally larger brains than most other mammals.

While the dawn horse's skull was slightly elongated, the face of the primate was shortened. The eyes of these primates had moved slightly forward on their faces, providing the beginnings of binocular vision. The ends of their forelimbs were no longer paw-like, but had become hand-like. The grasping thumb of which we humans are so proud was already evolving and appeared in very primitive form. It's not as if the future of the primate was sealed—evolution still had many cards to play—but we can easily see in these early primate skeletons our own ancestors.

Although the *earliest known* horse fossil comes from Polecat Bench, the first *named* dawn horse fossil was found in England in the 1830s. Ironically, the fossil was found by Richard Owen, a respected scientist who would become an outspoken critic of Charles Darwin. Owen did not recognize the fossil he described as an early horse. Instead, he named it *Hyracotherium* and suggested it might have been distantly related to the modern rabbit.

Because of Owen's misinterpretation, when numerous early horse fossils, recognized as such, started turning up in North America, scientists gave them a different name—Eohippus. The connection between the North American horses and Owen's *Hyracotherium* was not made for quite a while, since cross-Atlantic communications were limited. The question of nomenclature has yet to be settled even today: some scientists believe that Eohippus and *Hyracotherium* are the same animal, while others do not. These disagreements point out how closely related so many of these animals were during the early Eocene. To mistake a twenty-first-century horse for a twenty-first-century rabbit is inconceivable, but in the early Eocene, the many groups of mammals had only begun to differentiate. Many—horses and rabbits, for example—look quite similar.

Additionally, scientists in the nineteenth century fought very bitter battles with each other over horse evolution in general. Paleontological disagreements could be nasty, and understanding the evolution of the horse was particularly vexing to European researchers. There were many horse fossils in European rocks (the later horses were more easily recognized than the dawn horses), but, oddly, these fossil horses only turned up in certain layers of time. In the Eocene epoch, which ended about 34 million years ago, Europeans found many small horses. Then, for quite a while, horses seemed to be absent from Europe. Rock layers from 10 million years ago yielded horses again. Lots of horses. And this time, they were much larger and unmistakably horselike, even though they still had three toes. Then, finally, one-toed horses turned up in great abundance.

The erratic behavior of horses over time gave Charles Darwin a serious headache. For Darwin, the process of evolution was gentle and even-paced, like a soft English summer rain. Darwin didn't much care for perturbation—when the stress was too great he retired to his favorite health spa—and his theory of change did not include sudden heat spikes or asteroid impacts or mammalian land grabs. These discoveries came long after his death. In Darwin's society, God would not make an undependable planet. The fact that he was claiming that life changed in a world that most other people believed to be completely static was enough controversy for him; the concept of sudden spurts of change was way beyond what he was prepared to think about.

And yet—here were these crazy horse fossils. There just seemed to be no continuity over time. Horses appeared and disappeared in Europe with an absolutely rude abruptness, presenting a kind of magical "now-you-see-'em, now-you-don't" aspect of which Darwin thoroughly disapproved. It wasn't just that the horses behaved oddly across the ages. It was that when they reappeared after long absences of millions of years, they were *different*.

Different not in just small ways, as the Sable Island horses have legs a bit different from the legs of horses who run on the open plain, but in seemingly huge ways. First they were small, then they were large. First they had four toes on their front feet, then suddenly three toes—and then only one toe. Darwin was thinking about all this long before we knew about genetics or DNA, and from the European point of view, the evolution of horses flew in the face of logic. It seemed downright flaky.

Even worse, the apparent lack of steady steps leading from one version of the horse to another provided fodder for his enemies, including Owen. Darwin and other researchers didn't know that there were large pieces of the horse's evolutionary puzzle yet to be discovered, and no one ever considered that those puzzles lay in the New World. When Europeans first sailed into the Western Hemisphere, there were no horses to be found. None at all. Not in North America. Not in South America. Not on the plains and not in the mountains. Scientists, who then knew nothing about plate tectonics and the ever-widening Atlantic Ocean, simply assumed that horses were Old World animals and had never lived in the Western Hemisphere.

As a young man, Darwin had found an important clue to this mystery, but he didn't entirely understand the clue's significance. During a several-year voyage on a British research ship, *The Beagle*, he had taken a side journey high into the Andes Mountains in Chile. There, of all places, he found a fossil tooth that clearly came from a horse. So what was the deal? How did that tooth get to the top of those mountains? He had no clear answer. In Chile, he had experienced an earthquake which, he could see for himself, raised the level of some land by several feet, and he did understand that mountains could be created by such forces. But what was a horse, seemingly an Old World animal, doing there?

Further piquing his curiosity was another odd clue: the horses brought to the New World in the 1500s by Europeans lived long and prospered—on both the South American pampas and on the North American plains. A few escaped horses burgeoned into a fabulously prolific Western Hemisphere population. In only a century, the hundreds of imported horses quickly numbered in the tens of thousands. They seemed to be perfectly at home.

Darwin was confused. All the evidence—(1) the horse tooth high in the Andes, (2) the plentitude of horses in certain layers of European rock but not in others, and (3) the absence of living horses in the Western Hemisphere before the explorers arrived, followed by their proliferation on the pampas and plains—gnawed at him. The world just wasn't supposed to be so unstable, but the horse fossils found in the European rock record *seemed* to tell a different story: Here today, gone tomorrow. It was one thing to tout a theory of soothingly logical long-term change and quite another to claim that the natural world could change rather quickly.

This was where things stood in 1877, when the English genius Thomas Henry Huxley, supporter of Charles Darwin and archenemy of Richard Owen, arrived in New York City to present a lecture. Prior to his presentation, Huxley went to Connecticut to visit the paleontologist O. C. Marsh at Yale University. A fanatical collector of fossils, Marsh had hundreds of ancient horse bones gleaned from the sands of Wyoming and elsewhere in the American West.

Marsh laid out for Huxley a whole progression of horse fossils, including many fossils that did not exist in Europe. Huxley saw exactly what Darwin had been yearning to see: Marsh's arrangement demonstrated that the horse's leg had indeed evolved over millions of years. Marsh showed Huxley examples of various steps along the way: horse fossils with four front toes, then a slightly later horse with three front toes, each nearly the same size, then a later horse still with three toes but with one very large middle toe and two smaller side toes, then an even later horse with a very large middle toe and two side toes so small that Marsh thought (mistakenly) that the toes were useless. Then, finally, there was a horse with only one toe.

Logic at last! Order restored to Darwin's universe! Huxley, elated, sent the headlines to Darwin. To Darwin, evolution was about the "ascent of life," and horse evolution now seemed to prove him correct. Horses had indeed started off as small and insignificant animals and then had, by seemingly slow and steady steps, become what they "should" be—magnificent and powerful. There was both organization and—that best of all possible Victorian nouns—progress.

The mistake the scientists had made was in assuming that horses had evolved in the Old World—a mistake that was reasonable, given that there were no horses alive in the New World when Europeans arrived. Marsh's research made clear that horse evolution had occurred primarily in North America, rather than in Europe. The episodic appearance of horses in the Old World rock record was due to the fact that only certain New World horse species managed to find their way over to the Old World.

Marsh was also delighted. As a student studying in Europe, he had been taught that "the horse was a gift from the Old World to the New." Now Marsh showed that the opposite was true. The horse, it turned out, was a gift *to* the Old World *from* the New. For him, it was a matter of hemispheric pride.

But wait, there's more, Marsh told Huxley during their meeting at Yale. Marsh had an early primate fossil—also found in the American West.

For Huxley, the truth dawned: horses and primates had been partners for much longer than anyone had realized. Huxley sketched a humorous cartoon that showed an imaginary Eohomo riding an imaginary Eohippus and gave it to Marsh.

Huxley's humorous depiction of Eohomo and Eohippus (Courtesy of the Peabody Museum of Natural History, Yale University, New Haven, Connecticut)

Long before I met Ransom and Preator and Kelley, pondering the surprising importance of horses to the science of evolution, I visited Chris Norris, the paleontologist who coined the phrase "asteroid porn." Norris is keeper of Marsh's horse bones at Yale's Peabody Museum of Natural History. I asked him why horses had always been at the center of the discussion of evolutionary innovation.

"To understand change," Norris answered, "you need a good fossil record—and horses are extremely abundant. It's possible to look at their history over time in ways that aren't necessarily possible with other animals."

You can also do this with seashells, he continued, but the story of shells lacks a certain panache. People aren't readily struck by the drama of changing shell shapes. The change in the number of toes on a horse's foot, on the other hand, is pretty easy to see.

"Shells can't tell you a compelling tale in the way that horse fossils can," Norris said. "Horses are iconic and accessible."

Then he added: "Horses can tell you a story."

Of course, the story as understood by Darwin, Marsh, and Huxley was only partly correct. For them, the change in the number of toes on a horse's foot represented a lovely Victorian tale of perfection. To them, horses "should" have had only one toe—one hoof per leg—and a jaw full of thoroughly efficient grinding teeth that allowed them to eat mouthful after mouthful of delicious hay.

The Victorians saw the early dawn horses as inevitably leading to their own modern, majestic animals. In their view, horses were first small and innocuous, then "progressed" over millions of years until the best of all possible horses had evolved to live in the best of all possible worlds with the best of all possible primates—*us*. For the Victorians, evolution was a very directed experience. The end result was foreordained. There was no meandering down paleontologically convoluted mazelike paths that eventually led to dead ends.

Today we know that the story told by horses is not a tale of journeying toward the "perfect," but about the miracle of change on a changing planet. In *Reading the Rocks*, the geologist Marcia Bjornerud explains this idea a bit more formally: "Natural systems are remarkably robust precisely because no regime is permanent and no equilibrium is absolute." The story of horses as we understand it today is every bit as good as Charles Darwin's story, but it's a story of *process* rather than of *progress*: The Sable Island horses have developed goatlike legs because of their unique living circumstances, and not because they are moving toward a particular, ultimate goatlike destiny. The tiny dawn horses changed over time not because bigger was better, or because faster was better, but because the world around them changed. Tectonic plates collided. Ocean currents shifted. Mountains rose. Mountains eroded. The world became hot. The world became cold. Given this kind of geological and meteorological pandemonium, had the dawn horses been "immutable"—not able to evolve—horses would have died out quite quickly.

Our planet seethes with energy. For life to survive, it must keep up with the planet's convulsions, both abrupt and gentle—and the thrilling truth is that we've been able to do this. Horses and humans are survivors. Polecat Bench has provided an outstanding and entirely unexpected example of this truth.

For years, Philip Gingerich held the record of having found the oldest known horse fossil, discovered at the bench. But recently, other paleontologists found a dawn horse fossil that was just a bit older than the fossil found by Gingerich. Interestingly, this fossil was not only older—but bigger. This slightly older and slightly bigger animal lived in a world that was also slightly more lush. Then, as the heat rose at the beginning of the sudden spike, the world of Polecat Bench dried up. Plant life responded by changing. Horses responded to the shift in plant life by becoming smaller. Gingerich told me his horse fossil was about the size of his Siamese cat. The older, more recently found fossil is about the size of a small dog.

Gingerich has other evidence that horses made changes according to the environment in which they lived. We know that horses were even then quite gregarious, since their fossils are often found in groups, but their social arrangements depended in part on whether they lived in thickly forested areas or in more open woodlands. Gingerich found that fossils of males and females living in open woodlands showed clear size differences, with the males being about 15 percent larger than the females. Dawn horses living in thickly forested regions were about the same size whether male or female. Gingerich concluded that mares living in open areas gathered into groups and that males had to fight each other in order to be close to these mare groups. Mares living in more thickly forested areas were more independent of each other, and males had more opportunities to mate without needing to fight. Therefore, a larger size was not necessary.

However, some traits are more flexible than others. The foot bones of the Sable Island horses were able to change rather quickly, but teeth, it seems, usually change rather slowly. When we watch horses today, we marvel at their hoofs, at their powerful necks, at their ability to stand on their hind legs. We rarely look at their teeth. But it's modern horses' teeth that make it possible for them to live in so many different places and to survive by eating gorse or ultra-sandy beach grass. Dawn horses would probably not have been able to survive for long on the food the Sable Island horses eat.

Modern horses have fabulous teeth, teeth that last a lifetime, teeth that are four and five inches long and that are buried so deeply in the jaw that most of us have only a vague idea of how long, powerful, and efficient they are. I was taught as a child, as were many of my equestrian friends, that horse teeth literally grow constantly over the horse's lifetime.

This is not the case. Just like us, horses have fully formed teeth by the time they are young adults. But while our teeth emerge fully in childhood and then (hopefully) stay pretty much where they are for a lifetime, horses' teeth do not. With horses, these fully formed adult teeth emerge from below the horse's gum line much more slowly than do ours. The process of *emergence* may sometimes last as long as twenty years for horses.

In free-roaming horses, this slow emergence is a blessing. It allows a horse to ingest sand on Sable Island or to eat food like gorse, both of which wear down teeth. If the teeth of free-roaming horses didn't constantly emerge, they would quickly wear out and the horses would die without reproducing.

Dawn horses didn't have huge teeth. They didn't need them. Their delicate muzzles were often pointed upward, and their front teeth, their upper and lower incisors, were efficient at nipping buds off branch tips. These noteworthy nipping teeth would stay with them and change as plant life changed for the next 56 million years.

Neither were the dawn horses' molars large. They didn't need to be. Instead, their molars were shaped for smooshing fresh fruits, rather than grinding them up. These teeth would change dramatically over the coming years, to accommodate new foods, such as all kinds of grass. The dawn horses would never have been able to make a living by grazing on grasses, but this didn't matter anyway because there were no grasslands, as it was still too wet for grasses—"drought specialists," according to the Saskatchewan author Candace Savage—to proliferate.

Still, for their day, the teeth of dawn horses and dawn primates were high-tech and full of subtleties. Most field paleontologists can look at a fossil tooth and know not only whether it comes from a mammal—but from which species.

Some can do this with only a fragment of a tooth—a skill for which I have a great deal of respect, as I lack the required patience. Countless paleontological papers have been written discussing the size, down to the micron level, of the various bumps and lumps on only one tooth.

This is because often a tooth is all a paleontologist might have to work with. Bones are fragile things, but teeth, hard and dense and already partly mineralized, last forever. Consequently, it's not unusual for a new species discovery to be based entirely on the identification of only one tooth. Nevertheless, spending a day reading a stack of scientific papers describing the dimensions of various mammal teeth can be a tedious business. One's eyes (*my* eyes) tend to glaze over.

"Teeth in and of themselves are not entirely fascinating," I said one day to Chris Norris, by way of backing into the issue without (hopefully) seeming rude. I'd made a similar comment to another paleontologist who'd said, with great indignation: "Hey, that's my life's work you're talking about." I'd tried not to offend this time around.

Norris, fortunately, agreed with me, admitting that he himself was not initially taken by the study of fossil teeth. But, he explained, there's a lot more information available in an individual tooth than people might realize.

This surprised me. I hadn't thought of teeth as providing any information at all. Another paleontologist, Mike Voorhies, explained it to me this way: "Teeth have a memory."

It turns out that teeth can provide all kinds of clues to an animal's diet and lifestyle. Compared to those of most reptiles, our mammal teeth are super-sophisticated. Most reptiles can only slice and dice. We mammals chew—grind, even—which allows us to eat tough stuff like raw carrots, which in turn allows more opportunities to eat. We gained the ability to get quick sugar fixes from foods like ripened fruit (thank you, Cretaceous Terrestrial Revolution), which in turn helped us grow larger brains. The teeth of Polecat Bench horses and primates show that we developed this fondness for quick energy at least 56 million years ago.

Thus began what paleontologists call the arms race between mammals and plants.

"So," I said to Norris, "teeth are where the rubber meets the road?"

"It has to do with being a mammal," Norris answered. Being warm-blooded is expensive, energy-wise. You have to get the most out of every bite you take. "Mammals have to be extremely efficient in their digestion. In mammals, that efficiency starts in the mouth. We're not like crocodiles, that just gulp things down. Mammals chew, and many mammals chew plants."

So it's a case of "Waste not, want not."

Plants, of course, then developed their own survival strategies. This explains why some of them, like gorse, evolved some pretty effective defensive measures.

"Plants don't have the option of escape," Norris said, "so what they have to do is defend themselves. For a plant, it significantly matters if it can stop you eating more than a few leaves of it. For mammals, the mouth is the front line of that battle."

It's a teeth-versus-plants knockdown, and evolution is always raising the stakes. As plants became more and more aggressive in adopting don't-eat-me strategies, many animals would fail to keep pace. Horses, however, would consistently rise to the occasion and turn out to be supreme combatants.

Horses, science is showing us, are natural early adopters.

Scientists suspect that the earliest horses and primates lived everywhere in North America, but there aren't many sites that directly prove this. They probably exist, but they are hard to find, and when they are found they are sometimes even harder to excavate. As a young man, the paleontologist Chris Beard (another Polecat Bench graduate) learned this firsthand. Beard found an early primate fossil in Yale's collection that was in a drawer marked "Mississippi."

This, Beard thought, was an error.

"I knew for a fact that there were no Eocene primates in Mississippi," he told me. All the textbooks said that. But, "lo and behold," he said, there it was.

Curious, Beard went hunting. And, lo and behold, he found Eocene primate and horse fossils in Mississippi—in a horrid place, prospecting-wise. Poison ivy. Sticky pines. Snakes. Mud. Kudzu. Hot. Humid.

"The worst place in the world to find fossils," Beard said. "Back-breaking labor," he also said. "We had to remove all of the overlying rock and get down to the productive layers. Then we used small tools to harvest this productive layer of rock."

Listening to Beard, I started to think of modern-day Polecat Bench as rather pleasant and resort-like.

"We used screen washing," he explained. "To the average person it looks like panning for gold. We take this sediment that has fossils in it. We run it through a fine-mesh window screen. Anything bigger than the mesh is left behind. The screen-washed concentrate or residue goes back to our lab at the museum. Technicians look under a microscope to try to separate the wheat from the chaff."

He did this during the spring and the fall for ten years. Finally, he was victorious: He found a fragment of one horse tooth. He also found an early primate fossil. "It turned out to be incredibly interesting from my perspective. The horse fossil is incredibly fragmentary. It's only part of a lower molar. If I showed you the fossil, you would be unimpressed. But it's completely diagnostic."

Beard likes adverbs.

I wondered what "completely diagnostic" meant.

"As a mammal, your teeth have an incredibly complicated topography, full of crests and valleys. For every species of mammal, this topography is distinctive—a fingerprint at a crime scene."

So there we were—primates accompanying horses down in Mississippi during the early Eocene—at just about the time we were enjoying life together in wet and wild Wyoming. We think of horses as grassland creatures, but, just like us, they apparently enjoyed a tropical lifestyle tens of millions of years ago. When I thought about it, I decided that the free-roaming modern horses I'd seen in wetlands around the world were just doing what they'd always done. Living on sea islands comes just as naturally to horses as it does to us.

3

THE GARDEN OF EDEN APPEARS, THEN VANISHES

> Living horses, with their high-crowned grinding teeth and single hooves, bear little resemblance to the first horses . . . poodle-sized creatures with four toes and low-crowned, bumpy teeth.
>
> —MICHAEL NOVACEK, *Dinosaurs of the Flaming Cliffs*

The wind is a crude excavator. It has all the finesse of a pickax. So although the winds of the American West have revealed lots of fossils, the finer details of an animal's life are usually lost.

If any early horses had mustaches like the Garranos studied by Laura Lagos, for example, those mustaches won't be preserved on Polecat Bench. We can find teeth. We can sometimes find bones. But the more delicate tissues are rare indeed. The Wyoming wind has seen to that.

This lack of detail is frustrating. In Wyoming we can learn almost nothing about the lifestyle of these earliest horses. How did they live? How many foals did the mares have? What did they eat? Hard facts, like soft tissues, are distressingly scarce. We suspect that the Wyoming horses ate fruit, because of the clues provided by the shapes of their teeth. We can *deduce*—but there's no direct evidence.

There is one special place in the world, though, where the secrets of Eocene life are revealed with spellbinding clarity. Even the totally uninitiated, like me, can understand the overwhelming importance of what's buried here. Standing on a small viewing platform placed on a hillside just south of Frankfurt, Germany, I looked down into a huge hole in the earth. I was at another fossil site that also once hosted a plethora of early horses and early primates. My guide was the paleontologist Stephan Schaal.

This site, called Grube Messel, dates to the Eocene epoch, just as does Polecat Bench. But where Polecat Bench is now windblown and desiccated, the German site is serene and lush, much as it was 47 million years ago. Its treasures are safely ensconced between layers of wet claylike material.

While Schaal and I stood and talked, gentle breezes rustled the leaves on the plentiful trees and shrubs. The early fall temperatures were cool but not cold. The air was crisp and invigorating. From my lookout point I could see below me a marshy area rich with plant life—about as different from the present-day world of Polecat Bench as I could imagine.

Down there, Schaal said, pointing to one of the world's most spectacular fossil sites, lay the graves of hundreds, probably even thousands of ancient mammals, including dawn horses and early primates. Also entombed at the site are hundreds of thousands (maybe millions) of Eocene insects along with countless species of vegetation.

At this World Heritage site Eocene life is floridly, gloriously, sumptuously present. Everything is there. From soup to nuts. Almost nowhere else in the world is such a complete ancient ecosystem so well-preserved.

When animals and insects died at Grube Messel 47 million years ago, almost 10 million years after the Polecat Bench horses and primates, they sank down into an anaerobic deep-water lake bed. In that oxygen-free environment, without bacteria to break down the animals' tissues, their bodies did not decompose. Instead, the flesh, feathers, ligaments, tendons, and skeletons lay intact, season after season, year after year, on through the ages. Layers of detritus drifted down through the pond water to blanket them until, finally, they were entombed in a mille-feuille pastry of ultrathin alternating layers of silt, then algae, then silt and algae again.

To peel back these layers is to turn, quite literally, the pages of an

Eocene Book of Life. Opening that book by separating the layers, we can easily see that the world of 47 million years ago was similar in many ways to the world of Polecat Bench. Despite the passing of almost 10 million years, the world was still wet and swelteringly tropical. The heat hadn't stayed high, though. The extreme temperature spike that began the Eocene and that, graphed out, reminded me of the Eiffel Tower, lasted only a few hundred thousand years. World temperatures, which had risen so markedly at the beginning of the Eocene, plunged almost as abruptly, finishing off the PETM. Then, temperatures again began to rise, only this time much more gradually. The graph of world temperatures across time now looks like a gentle hill rather than a spike.

Again, the cause of the gradual increase is uncertain. Evidence from deep ocean cores shows that the planet's atmosphere was then filled with greenhouse gases, which resulted in the slow warming, so scientists understand that greenhouse gases brought about the temperature increase—but no one knows where those greenhouse gases came from.

In any case, by the time the carcasses of horses and primates and the other plants and animals were folded so gently into the clays of Messel, the heat was once again back at early Eocene heights. There was no ice in the world, neither at the poles nor on the mountain peaks. Sea levels were very—*very*—high, so much so that much of what we know today as Europe was then under salt water. What land was above sea level consisted of islands. From Grube Messel, it was a long swim to Asia.

Nevertheless, the European life isolated on those islands was rich and plentiful. Inside the layers of Messel, like keepsake flowers pressed in between a book's pages, horses and primates are accompanied by a secret garden of flora and fauna, some of which we still have with us in the world today.

Schaal and I walked down the steep hillside to the pond and wetland where a paleontological crew was carefully peeling apart the layers of silt and algae. Mining the fossils here is easy and fun: lots of immediate gratification, no backbreaking labor, no risk of sunstroke, no tramping through kudzu and poison ivy, and, perhaps most important, no working for ten years to find one tooth.

Invited to try, I found that I could sometimes open the layers with just my thumbnail. No need for hammer, chisel, or shovel. The task simply requires some delicacy. You have to be careful, because the pages of this manuscript are exceptionally frail. I didn't have to don white gloves, as in research libraries, but the thrill was just as intense.

Quarrying for fossils at Messel is, in the initial stage, a bit like quarrying stone: someone cuts a large chunk, a "book" of fossil-containing material several feet square and several feet deep, out of the layers of solidified mud, then carries it up from the pit to where people are working. Then comes the easier, more exciting task of separating the layers.

I walked from the work site down into the quarry itself, to feel the material in situ, with my own fingertips, where the fossils had lain for all those ages.

"Be careful," Schaal cautioned. It had begun to rain heavily. The footing going down into the pit was slippery. A lot of people fall.

I nodded. But his warning didn't register. I was too excited.

My feet slid out from under me. I felt with my fingers a "rock" that was not at all firm, but slightly pliable. This was not rock at all, although researchers sometimes use that term. It felt smooth and soft, like unbaked pottery clay.

Fossil preservation here is so magnificent that details like the hairs on a horse's tail or the teeth of a foal still inside the mare are easy to see. Even the colors of an insect's wings—formed not by pigment but by wing-surface striations that reflect light like the ridges on a CD—are still visible. A jeweled beetle fossil still shows a shiny metallic blue, just as it did when the insect flew through Eocene forests. Bird feather fossils show not only the feather's shaft and branching, but even the tiny individual hooks at the end of a feather's individual barb. The details of the scales of a huge crocodile remain clear. Small moths have been found half digested in bat stomachs. Fish bones have been found in the gut of an ancestral relative of the modern hedgehog. Pollen grains are still visible in the pollen sacs of flowers. The flower of an ancient water lily looks very much like a water lily looks today. The delicate bones of the inner ear of various bat species show that, even then, bats had perfected their ability to echolocate. For some plants, we can see not just the roots, but the individual rootlets.

And just as primates accompanied horses at Polecat Bench, primates are also present at Messel. One, nicknamed Ida by scientists after she was discovered at the end of the twentieth century, is fully articulated and is said to have a *Hautschatten*—German for "skin shadow." We can see the complete outline of her body, showing how the skin and flesh covered her bones. We know her long tail was covered by fur because we can still see the individual hairs. Ida is small, not even a foot long. She still has her baby teeth. We can also see her adult teeth that were forming in her jaw when she died. She enjoyed a final meal of leaves and fruits, which we can see partially digested in her stomach. Her hands are gnarled up in deathly rigor mortis. Like us, she has five fingers. Like ours, the small bones at the tips of her fingers are clearly shaped to be protected by nails rather than by claws or hoofs. Ida is so well-preserved that researchers can see that she had broken her wrist, but that the break had healed. Ida, aka *Darwinius*, was for a time at the center of a minor controversy, as some paleontologists suggested she was directly related to the primate line that led to *Homo sapiens.* The current consensus is that she had already separated from that particular group of primates, but only just. The differences between Ida's line and our own, at that point in time, are minor.

Among Ida's companions are several horse species, including a horse with four toes on the front feet and three on the back, closely (but not precisely) resembling the dawn horses of Polecat Bench. So the little dawn horses were, at least for a time during the Eocene, spread far and wide throughout the Northern Hemisphere.

In Wyoming, researchers find few fully articulated remains, but at Messel, questions about how the dawn horses spent their days are answered: we can see the animals in eerie detail, almost as if they had been X-rayed. The detail provided by some of the horse fossils is as complete as the detail provided by Ida. Just as we can see evidence of Ida's last meal, gut contents in one of the horse fossils show that the animal died with grape seeds still in the digestive system. In 1975, the German paleontologist Jens Franzen unearthed a complete Messel fossil of a young male *Eurohippus*, with a skin shadow showing a detailed outline of the animal's fleshy parts and with a gut showing partially digested plant matter. Later, other well-preserved *Eurohippus* were found. One fossil preserved the skin shadow of the little horse's deerlike outer ear, not yet evolved into the pointed ear of the modern horse. Another showed the horse's stubby tail, complete with the outlines of coarse tail hair. The tail would have accomplished nothing in the way of dealing with horseflies.

At Messel the horses matured quickly. Researchers have found a foal of about three months (they can tell by the maturation of the bones) that's already two-thirds the size of an adult. Some of the pregnant mares are themselves so young that they still have their own baby teeth. Reproducing at such an early age is unusual in modern horses, but it may well have been normal then, suggests Franzen. Since these horses lived for only a few years at most, this ability to reproduce quickly is probably one reason why they were so prolific. Messel also reveals other biological features that made the horse resilient enough to endure through the ages. One of them, surprisingly, is the horse's unique digestive system.

When I was taking care of Whisper, I found myself fascinated by his strange eating preferences. In the barns where I rode as a child, horses were fed a small bit of grain and a flake of hay in the morning, and ditto in the evening. That seemed to suffice. But Whisper quickly taught me that he wanted something different: he wanted to head over to a certain corner of the pasture for a bite of one thing, then walk somewhere else for another choice blade. He wasn't persnickety: he never turned down good food. But left to his own devices, his food choices were extremely varied. This kind of casual on-again, off-again snacking throughout the day is typical of horses on their own. It's "throughput" eating, different from the way cud-chewing cows, for example, eat.

Messel shows us that the difference between the grazing style of horses and that of cows has its roots in deep time. Even-toed animals like cows are ruminants who digest and redigest their food in a series of stomachs. This is why cows are said to chew their cud and spend roughly half their time in a pasture ruminating rather than grazing. Cows are patient food processors.

Horses are not. They spend much less time digesting each bite they take and a lot more time taking bites wherever they can find them. Their constantly grinding teeth and high-speed digestive system extract whatever energy is there. Our own stomach, and the four stomachs of cows, are of prime importance when it comes to digestion. But for the horse, the stomach is merely one stop—and not necessarily the most important stop—along the digestive tract. A carrot is digested in the mouth, in the stomach, in the intestines, and also in the cecum.

This cecum is an important key to the evolution of the horse and an important reason why his eating habits are so flexible. The cecum is large—quite large—and contains bacteria that break down food like gorse into nutrients that the horse can use. It's a major reason why horses can survive on a high-fiber, low-protein diet, and it's an important reason why horses, over 56 million years, were more adept than many other animals at eating newly evolved plants.

Many mammals have a cecum. We do. It's a small pouch that sits between the small and large intestines. The modern horse cecum, on the other hand, is four feet long. Without this organ, essentially a fermentation vat, horses would not be able to live in so many different ecosystems or eat such a variety of foods. The horse's digestive system is like a conveyor belt where food moves in at one end, passes through at comparatively high speed, and comes out the other.

The cecum houses combinations of bacteria, protozoa, and fungi that cooperate to break down cellulose. These digestion-facilitating communities are not the same from horse to horse, nor do they stay the same in any one horse throughout the animal's life. Instead, the composition of the community in a horse's cecum changes according to the diet the horse is eating. This change, however, must occur gradually, over the course of several days. Horses who change their diet too quickly may suffer greatly or even die. This is why you can't turn a horse out on spring grass for too long early in the season. The microbes in the cecum become imbalanced by a sudden, rich nutrient influx. Even altering the quality of the hay a horse eats can sometimes cause problems if the shift is too sudden.

Barn managers and horse owners often find this frustrating, but because of Messel, we know that an enlarged cecum as a solution to the problem of digestion is a solution with a deep history: as far back as the middle Eocene, Messel proves, horses could survive on forage that other animals would have to pass up.

While Schaal and I watched his team in action, I saw within the course of only a few hours the discovery of several complete specimens. One man shucked open a rock layer to discover an insect with two long, delicate antennae, each ending in a small sphere. Another peeled open a layer to find a fossil fish. He glanced at it and threw it over his shoulder into the scrap pile.

"Hey," I said instinctively, forgetting my manners. How is it possible to throw away something so precious?

He explained that the Senckenberg Museum in Frankfurt, which houses many of the fossils, already had so many of those fossilized fish that it wasn't worth preserving more. In fact, there are so many fossils at this site that the problem is not finding them, but preserving and storing them.

I saw this firsthand when Schaal took me to the Senckenberg and offered to show me one of his prize horses. On an upper floor of the museum, he pulled open one of the many cabinet drawers that contain preserved fossils. Inside this one was a tiny mare carrying an even tinier partially developed foal—with developing teeth. The baby lies "backward" in the uterus, as do today's foals.

He pulled open another drawer and out came another horse fossil. But this one looked strange. He explained that a chunk of the fossil's backbone had been removed by the business end of a bucket loader that had been digging up the Messel pit in order to make a garbage dump.

I did a double take.

"How did that come about?" I asked.

In the 1980s, the German government decided to convert Grube Messel—then not yet a World Heritage site—into a national garbage pit. To many government officials, a big empty pit seemed like a great place to stash the modern world's embarrassing detritus of plastic bags and bottles.

In their defense, Messel did not at that point appear to be a prime research site. It was taken for granted. Local people had for centuries mined the pit for coal. Villagers knew that strange animals, animals no longer present on Earth, could sometimes be found pressed in between the clay layers, but those ancient fossils were not a priority when Germany, like Britain and the United States, found itself in the throes of industrialization and needed all the energy it could get.

Moreover, preserving the fossils was difficult: The same friable wet clay on which I had slipped and which had preserved the skin shadows of ancient animals dried and turned to dust almost as soon as it was dug up. When the clay dried up, the fossils disintegrated. The irony was haunting, sadly reminiscent of the Greek tale of Orpheus and Eurydice: the mere act of looking at the fossils—just bringing the fossils into the light of day—destroyed them.

Without adequate preservation techniques, there seemed to be little point in mining the fossils themselves, but some nineteenth-century researchers did realize that the site contained important fossils and worked hard to get them noticed. In 1875, one Rudolph Ludwig found a fossilized crocodile, putting Messel on the paleontological map. A subsequent paleontological survey affirmed the site's value, but then World War I and World War II intervened. The site was operated as an energy resource until Allied bombs destroyed its infrastructure. After the war, the mine was not reopened, and German officials were stuck with a useless hole in the ground.

Without the ability to preserve the fossils, which were by then recognized as significant, there seemed to be no future for the site. Finally, researchers found an artificial resin that, applied with the correct technique, solved the problem. Painted quickly after the fossils were removed from the pit, the resin held the bones in place.

Messel could now be legitimately thought of as a fossil hunter's "El Dorado," in the words of Schaal. Nevertheless, the garbage dump plan moved ahead. Bucket loaders began gouging away the layers of clay. Scientists ran ahead trying desperately to preserve what they could. Schaal, then a young researcher, was in front of one of these machines when it scooped up some clay and exposed a skeleton, which Schaal managed to save. The fossil turned out to be a complete horse (or would have been complete, save for that chunk of backbone removed by the bucket loader). This very fossil was what Schaal had shown me in the museum drawer.

This dramatic event drew attention to the problem. As the public learned more about the importance of the site, protests grew, and in 1990 the government abandoned the garbage-dump plan. Five years later, Messel became a World Heritage site. One of the horse fossils even appeared on a German postage stamp.

Messel today is one of Germany's most important *Lagerstätten*. *Lagerstätten* is the plural. *Lagerstätte*, the singular, is one of those long German words that require a whole lot of English words to explain. Basically, it means "a place where fossils are so outrageously well-preserved that scientists will never know all there is to know about the fossils there."

The Senckenberg Research Institute and Natural History Museum is the 1815 inspiration of the poet Johann Wolfgang von Goethe, an early proponent of science for the public. Prior to that time, collectors were wealthy men who kept their natural history treasures in their own homes, where only a few invited guests could see them. These collections, expensive to obtain, were symbols of status. Early collectors displayed stuffed animals from all over the world, had personal greenhouses in which unusual or even bizarre plants grew, and kept rock collections, coral collections, and just about anything else from the world of nature that the wealthy could think of to amass. If you were invited to dinner, you were expected to view (and to be impressed by) the proprietor's hoard of natural history objects, and, by implication, his powerful global reach. Goethe believed in the democratization of science—that these collections should be available to everyone. Many citizens of Frankfurt agreed with him.

Today the Senckenberg Museum remains much as it was during Goethe's era, filled with rows of cases containing shells and birds and corals for visitors to contemplate. There's very little explanation. But in the museum's basement is a modern exhibit, full of information that focuses on Messel and its horses.

While there, I watched a short animation that presented one scientist's view of how a dawn horse might have moved. In the animation, fossilized bones rise out of their 47-million-year-old bed of clay and assemble themselves into a complete dawn horse. The poor little fellow ambles forward in an ungainly gait that's anything but horselike in the modern sense. It's not quite a walk and not quite a trot. It's something else entirely, something you'll never see a modern horse do.

Intrigued, I called the scientist behind the animation, Martin Fischer of Friedrich Schiller University in the former East Germany. An evolutionary biologist and specialist in animal locomotion, Fischer had created the animation more than a decade earlier.

"If we did this today, I would actually change some things. I would lower the body in between the limbs," he said. "This horse should move more like a fox."

To me, I said, the animal looked like it was crawling.

"It should be even *more* crawling, because this poor animal was not a horse at all and had no idea that it would become a horse," he said. "Small mammals of this size all have a curved spine."

I thought of my border collie and his emphatically flexible backbone that can easily undulate not just up and down, but from side to side. Were the dawn horses like that?

Somewhat, he answered.

He continued: "The straight spine which you know for the horse is a very late acquisition from evolution. If you look at a galloping horse, you will see that the galloping comes out of the lower spine, the last seven vertebrae. It's very late in evolution that this occurs and you have to be very high up so that you can gallop from your legs."

Small animals, he said, would not gallop like this.

This was interesting. It's certainly true that larger horses gallop more smoothly than smaller horses, but I hadn't thought of the gallop as an artifact of height. It helped explain why even now, when I think about some of the ponies I rode as a child, my spine aches.

"Horses are actually the only dorsal-stable animal we have. That's why we can ride them," Fischer continued.

This sounded very convenient for us humans.

"Think of a cow," he suggested. "The spine movement is so tremendous that you cannot sit on the cow. That's why your American cowboys can only stay on a cow for five seconds."

I mulled that over. It is *possible* to sit on a cow or a camel or even an elephant, I thought, but Fischer's right that they're not very comfortable. I've never ridden a cow, but I have ridden both camels and elephants, and each time I realized how much I missed my horse.

I asked what *he* thought the little dawn horses would have looked like when they ran.

"Think of a fox or dog of the same size and you're very close to how they moved. They could gallop, but not like a modern horse. They had to gallop in their back."

I explained that I had always felt sorry for these first horses, with what I imagined to be a cumbersome gait.

"You have this horse-centered view," he answered. "You have to get out of that to see what these early horses moved like. In today's horses we have a limb-dominated gait. In the early horse, we have limb-dominated gaits in the walk and trot. But to go faster, they had to bring in the backbone."

This explained why Margery Coombs had told me that the dawn horses scampered. The gallop as we know it lay well in the future.

Modern horses, Fischer explained, have "this dorsal-stable locomotion." By this he meant that relative to other living mammals, modern horses have a firm but flexible back shaped in a way that allows a rider to find a firm, well-balanced, easy-to-maintain seat.

"This is a very peculiar thing. Particular to horses," said Fischer. "And it's the prerequisite to riding."

Because of the structure of the modern horse, he added, the best point of balance on a horse is at the withers, where the neck meets the back. I mentioned that when I visited Mongolia, I saw that riders stand in their saddles at the gallop, hovering over the withers. One of the reasons for this, Mongolians had told me, was that it freed up the horse's legs and tired him out less. It took a while for modern flat-racing jockeys to figure this out. For the longest time, jump jockeys in steeplechase races also used to sit down firmly on the horse's loins, but Mongolians have understood the best way to ride a fast horse over lots of ground for centuries.

"People in Mongolia ride on the withers of the horse because they know it is even more stable. When you ride up on the shoulder, you ride on the center of gravity. This is the point where it's absolutely stable and you have the least amount of any kind of motion in any direction," Fischer explained.

I thought of Huxley's cartoon of Eohomo riding Eohippus: Riding horses during the Eocene would have been rather difficult. Huxley's little primate rider would most likely have gone right over the horse's neck. The equine skeleton that would permit elegant postures and modern dressage wouldn't come for another 40 million years or so.

And long before that happened, yet another worldwide shift in temperature and plant life would help bring the little horses dangerously close to worldwide extinction. Throughout the early and middle Eocene, the different horse species were quite similar to each other. Horses evolved, but in a rather cautious and conservative manner. While the horses of Messel enjoyed their island-living lifestyle, the horses in North America were changing.

Several hundred miles south of Polecat Bench at a place called Grizzly Buttes, a horse with slightly longer legs showed up only a few million years after the Polecat Bench horses first appeared. This horse, *Orohippus*, still had four toes on his front feet and he still had tiny teeth that could smoosh grapes, but *Orohippus* would have had difficulty with coarser foods. He certainly wouldn't have been able to eat anything like gorse.

Marsh, with his Victorian preoccupation with improvement, thought of this slightly younger and slightly larger horse as "better" than the first horses—a "new and improved" version, if you will. But now we know that *Orohippus* wasn't necessarily "better." He was just suited to a slightly different natural setting. In fact, for most of the Eocene, horses stuck with the same basic body plan.

They seem to have been one of evolution's great ideas.

Then, unexpectedly, they weren't.

Worldwide, horses nearly became extinct.

Several months after visiting Messel and the Senckenberg, I watched the paleontologist Matthew Mihlbachler unlock and pull open a small drawer sequestered in a distant corner of the monumental horse bone room of New York City's American Museum of Natural History. During the early twentieth century, the horse was so iconic that the

Museum of Natural History's logo, in true partnership style, consisted of a skeleton of a horse rearing beside a skeleton of a man. In the logo, the man's arms reach to the sky, just as do the horse's front legs, making the parallel between the modern horse skeleton and the modern human skeleton crystal clear.

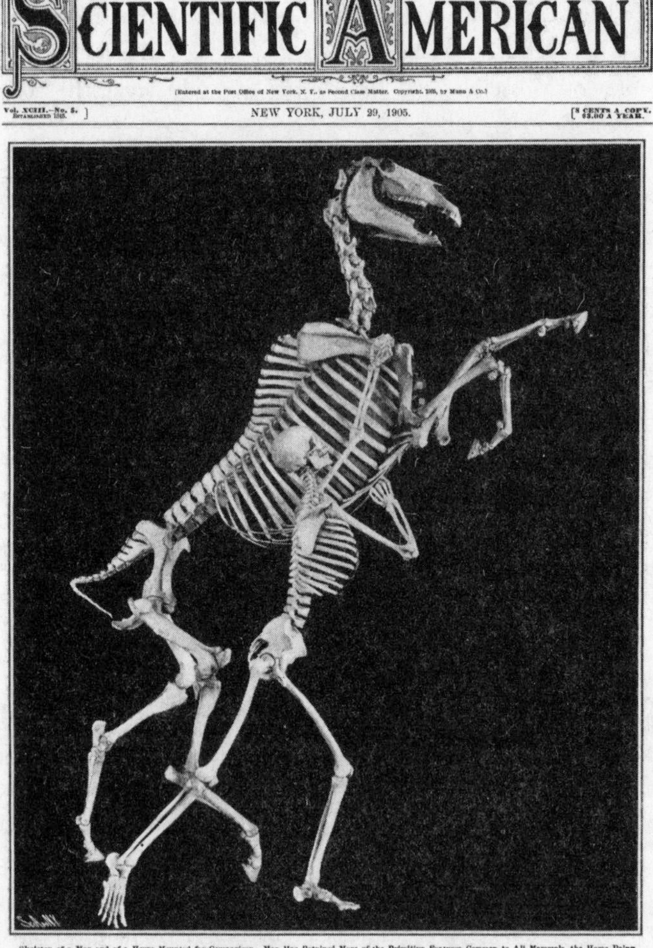

SCIENTIFIC AMERICAN

[Entered at the Post Office of New York, N. Y., as Second Class Matter. Copyright 1905, by Munn & Co.]

Vol. XCIII.—No. 5. Established 1845.] NEW YORK, JULY 29, 1905. [8 CENTS A COPY. $3.00 A YEAR.

Skeleton of a Man and of a Horse Mounted for Comparison. Man Has Retained More of the Primitive Features Common to All Mammals, the Horse Being Far More Specialized in the Structure of Its Limbs and of Its Grinding Teeth.

THE EVOLUTION OF THE HORSE.—[See page 81.]

The inspiration for the former logo of the American Museum of Natural History

"Look at this place. There is cabinet upon cabinet of stuff," said Mihlbachler, waving his hand to indicate horse bones covering the whole of horse evolution, from the days of Polecat Bench well into the Ice Age. The bones extended from one end of the cavernous room to the other.

"There's all this," he said, motioning toward endless rows of cabinets upon cabinets containing fossils of dawn horses. "And then there's this." He pointed to a few pitiful tiny bones in the corner of the drawer he had just opened: "Here's the middle Eocene of horse evolution in this one . . . very . . . empty . . . cabinet," he said. The drawer was labeled "*Epihippus*." "There's this huge bottleneck in horse evolution," explained Mihlbachler. "You're looking at it right here."

There was almost nothing in the drawer, save for a few tiny jaws and various fragments of bone and teeth. Museum researchers have been collecting horse bones for roughly 150 years and have cataloged tens of thousands of horse fossils. Plenty more are still in their plaster jackets, waiting for someone with the time and energy to open up the packages and find the treasures inside. So many fossils, so little time . . .

And yet, of all these gathered bones, almost none come from *Epihippus*, the mysterious horse of the late-middle Eocene, the horse that could have marked not the end of the beginning of horses—but the end of horses forever.

Epihippus fossils are as rare as early Eocene horse fossils are plentiful. I picked up a tooth, only a few millimeters in size. I mulled over what Mihlbachler had said. Had *Epihippus* become extinct, there would have been no Vogelherd horse, no horses to carry people over the North American sea of grass, no horses to bring Genghis Khan and his men from the far end of Asia to the gates of Vienna, no warhorses, no plow horses, no cow ponies or wild horses. We humans would be living a much lonelier life. Not to mention what might have happened to civilization itself.

"So this was it? Hanging on by a thread?" I asked.

Little *Epihippus* lived from about 46 million to 38 million years ago in North America. He turns up in the American West in a rock stratum called the Uinta Formation. Despite the countless numbers of horses who had covered the land during the early Eocene, toward the end of *Epihippus*'s time on Earth, horses seem to have become rare.

"We have drawer upon drawer of Uinta fossils in this museum, but that's all we have of *Epihippus*," Mihlbachler said. "It's hard to know why. They may have been superabundant in Ohio, but we don't have fossils in Ohio. They may have been supersuccessful somewhere else, but at least where we have fossils, *Epihippus* is just not common."

I had read about *Epihippus* once before, in Tim Flannery's wonderful epic of North America, *The Eternal Frontier*. "One wonders," the author had written, "how different human history would be if *Epihippus* had not made its knife-edge escape from extinction."

The basic problem appears to lie in the fact that early horses had not yet developed their evolutionary facility for adaptation. They were well suited to their heavenly Eocene hothouse. The foliage was lush. Food was abundant and available year-round. What's not to like?

But nothing lasts forever. By the time of *Epihippus*, the planet was once again on the brink of major change. The warmth was about to disappear. In its place would come a colder and much more challenging world.

For what may be the first time in Earth's history, "seasonality" would begin. All life would have to adapt to this shift. The year-round empyrean feast of ripe fruits and tender buds ended. For many animals, finding food would become burdensome. Birds would have to learn to fly between the Northern and Southern Hemispheres twice a year. Whales would have to learn how to migrate through the world's oceans on a seasonal basis. Bears would have to learn how to hibernate. Browsing mammals would have to learn how to stock up on fat during the warm season to carry them through the cold winters.

It's hard for us to imagine in our twenty-first-century world, where we accept the cruel reality of sweltering summers and freezing winters, but for a good deal of Earth's history, including most of the Eocene, the planet enjoyed fairly uniform temperatures. For example, during the Eocene, the world north of the Arctic Circle was so warm that crocodiles flourished there. There were no ice caps, of course, and so much freshwater flowed into the Arctic Ocean that a layer of freshwater sat like a lens over the salt water. The freshwater *Azolla* fern was plentiful. Forests of redwoods and walnut trees grew there. Paleontologists have found the remains of giant ants usually associated with the tropics.

Of course, the planet was still tilted and the poles still experienced long nights and long days during the solstice periods, but the warm environment was apparently so alluring that the animals made peace with the dark periods and lived there throughout the year. Fossils of primates and tapirs have turned up from this period in the Arctic rocks, along with snakes and alligators and turtles galore. Curiously, though, no early horse fossils have yet been found.

Then, toward the end of the Eocene, that frenzy of animal life was challenged. Their Elysian lifestyle came to a rather abrupt end. Consider that the fabulous and whimsical brontotheres, kissing cousins to the dawn horses, had ranged almost worldwide during the Eocene. Like the dawn horses, they had four toes on their front feet and three on their back feet. But unlike the dawn horses, they had taken great evolutionary risks. The first brontotheres were small, but then they exploded in size. Some grew to enormous heights, as much as eight feet at the withers. They were highly experimental. This risk-taking would prove to be their undoing.

This presents an interesting mystery: brontotheres and dawn horses appeared on Earth at about the same time, as far as we can tell, and are closely related, but while the horses remained evolutionarily conservative, brontotheres tried every evolutionary trick in the book. For example, they fooled around a lot with what are commonly called "horns" on their heads. These protuberances sit at the end of the skull, like rhino horns, but they are not horns at all. Horns are made of keratin, like our fingernails. The "horns" of the brontotheres are actually extensions of the skull. (Many animals evolve such skull ornaments, but, as Mike Voorhies had said, horses never did.)

In the short run, the brontotheres' evolutionary strategy of experimentation paid off: they were among the most diverse groups of land mammals on the planet, and by the end of the Eocene a few species were as large as modern elephants, making them the largest of land mammals then extant. At one point there were so many different species of brontotheres roaming the planet that Matthew Mihlbachler once wrote that trying to keep their names straight was a "dizzying" task.

But despite this proliferation, by the end of the epoch they were gone. Perhaps their strategy of becoming highly specialized meant that eventually they became *too* specialized. When their favorite ecosystems disappeared, it was as though they'd had the rug pulled out from under them.

Also failing to make the shift from a warm to a cold world were North American primates. Fortunately for us, though, by the end of the Eocene they'd made their way into the heart of Africa, where they survived.

Horses, too, were severely challenged. Used to a life of leisure, where the grapes were there for the taking, horses had to develop coping strategies. Their agile ability to scamper from one hiding place to another underneath the copious canopy would no longer suffice as North America's junglelike ecosystems evolved into more-open woodlands. In place of agility, the horses would need to opt for even greater speed as a defense. Even more important, their teeth would have to be able to grind rather than smoosh, as horses resigned themselves to eating much tougher forage.

Happily, just in the nick of time, just before the final, catastrophic end of the Eocene, a new and different horse appeared. *Mesohippus* was taller, faster, had only three toes on his front feet, and had teeth that were larger and flatter, with greater surface area. The improved ability to grind food was a great advantage.

And then, soon after *Mesohippus* came *Miohippus*, an extra-horsepower horse, even bigger and even tougher. *Miohippus* was a horse that most of us, at last, would recognize as a horse. He still had three toes, but his middle toe was clearly the weight-bearing toe. His face and skull were horselike and his backbone, while not like the backbone of the modern horse, had straightened out enough to allow him to somewhat gallop, rather than only "scamper."

After many millions of years of evolutionary cautiousness, horses were on the move. "Just in the nick of time" is not an exaggeration.

World temperatures had been slowly dropping, but then the planet suddenly plunged into a deep freeze. The precipitous drop was every bit as abrupt as the early Eocene's temperature increase. This sudden temperature change, which occurred about 34 million years ago, brought a cascade of consequences.

In Europe the cold caused an extinction event known as *La Grande Coupure*. "The Great Cut." As I mentioned, Europe during the Eocene was an archipelago of islands. When the temperature dropped, newly formed glaciers froze colossal amounts of the planet's water. Sea levels dropped and these islands became connected. The animals, long accustomed to the privacy and comfort of peaceful island life—disappeared.

In their place a new group of animals arrived. This was one of the strange, inexplicable anomalies in the European rock record that caused Darwin's emotional crises. There was a line of demarcation in the rock record. Below this line lie all the many different European Eocene mammals; above it are found very different animals—animals that had been common in Asia. This line is so clear and so obvious that early European paleontologists could *see* it, although they couldn't explain it. Having no idea that the world had abruptly plunged into a deep cold spell, Charles Darwin along with other scientists agonized over the *Coupure*, which seemed to contradict his slow-and-steady thinking about evolutionary change.

Today, though, we have a pretty good idea of at least some of the details. The *Coupure* was a perfect-storm incident involving deep-sea phenomena and plate tectonics. Researchers postulate a one-two punch of events, one event that occurred gradually and another that occurred comparatively suddenly. First there seems to have been a general decrease in the atmospheric gases that kept the planet warm. Some paleogeologists suggest that the tectonic plate of India, long parted from Africa and slowly drifting north, collided with Asia and began forcing the rise of the Himalayas, which in turn slowly cooled the planet by depleting carbon and oxygen from the atmosphere. Areas thick with foliage, like Messel, were replaced by more-open areas.

Coupled with this was the final coup de grâce: the tectonic independence of Antarctica. Just as India had broken free from Africa and headed toward Asia, and just as the North American plate was separating from the European plate and opening up the Atlantic Ocean, Antarctica had been slowly separating from South America and from Australia. Finally, about 34 million years ago, Antarctica severed all physical contact with the other continents. It sat, by itself, right over the South Pole. Scientists suggest that this event had several profound effects.

First, Antarctica grew an ice cap and became a global Ice Queen who ruled the world with a frigid fist. So much water was tied up in the ice that Europe was no longer a series of isolated islands, but an expanse of above-sea-level land connected to Asia. As such, Europe was open for conquest. The Asian animals, better suited to the new conditions, rushed in.

North America, too, experienced changes. Sea levels dropped, exposing former seafloor to colonizing plants. Because so much water was tied up in glaciers and ice caps, expanses of grass appeared for the first time. Animals like *Miohippus*, capable of living on those season-dominated, drier, open plains, also appeared.

The planet's ocean currents shifted. After Antarctica came to power and was comfortably ensconced in her South Pole castle, she was surrounded by a circumpolar current that acted like a moat. This current fed cold water into the world's other ocean currents, changing their flow and bringing about a colder planet.

Chris Norris had said that horses had a unique story to tell, and it turned out that their story was one of adaptability—but an adaptability that depended on context. Horses were born into a silver-spoon existence, into a world where food was easily available and temperatures were comfortable. Think *Porgy and Bess*: "Summertime, and the livin' is easy."

Then the world changed, and the horses were tested and hardened. Fortunately, they did manage to evolve enough to hang on and pass into the chilly, challenging, brave new world of the Oligocene. Their unusual digestive systems and their longer legs helped them, but it was their teeth, according to Matthew Mihlbachler, that did the heavy lifting.

Mihlbachler once pulled more than seven thousand North American horse teeth out of a multitude of drawers and cabinets at the American Museum of Natural History, hoping to correlate horse evolution with climate change. He unlocked and relocked cabinet after cabinet.

"This is by far the world's largest collection of fossil horses," he said. "If you go from one end to the other of this floor, you'll witness the entire evolution of horses. We opened up every drawer on the floor and looked at the upper teeth of every fossil horse in that collection."

He is a patient man.

"We looked at the cusps to see if they were worn or flat. We got a huge spreadsheet going. We also went to Yale. There's a few gaps in our collection that Yale fills nicely. We took those results and we converted the cusp sharpness into a scale. We mapped all that tooth wear data through time and compared it with paleoclimate data."

Mihlbachler found evidence that following the 34-million-year-old temperature drop, horses changed their diet by shifting from eating fruits to eating other forage—forage that gradually became more and more abrasive and that grew close to the ground, so that horses would ingest a lot of grit along with their grass. Some horse species had teeth that were able to cope with this shift. Those who did not, disappeared.

Those who survived had tougher teeth. If you've ever been to a beach and accidentally gotten a mouthful of sand, you'll understand the problem that horses encountered when they began to graze rather than browse. Learning to eat grass—a formidable food that defended itself from predators like horses by incorporating more and more razor-sharp silica into its blades—was probably not a pleasant task. Silica is the same material from which glass is made. If you've ever walked through a field, picked a blade of grass, and cut yourself by running your finger along the fine edge of the blade, then you'll understand the problem the horses faced. Your skin has been sliced by what's essentially a sharply honed knife.

Silica can be dangerous stuff. We humans can chew on a blade or two of grass over the course of a day, but we would never be able to survive on the stuff. For horses, gaining the ability to eat grass was a supreme achievement. In his study of the seven thousand horse teeth, Mihlbachler saw that colder temperatures, plant shifts, dietary changes, changes in silica content, and changes in horse teeth were all correlated. It was as if horses and grasses were waging a fairly constant war.

Interestingly, other animals were not able to respond this way. Mihlbachler looked at the teeth of camels, which were also plentiful in North American rock layers. But camels, it seems, did not have the same talent. Their teeth did not change as much.

"Horses are really tracking global temperatures, according to their teeth," he said. "The horses just kind of change with the times. I don't know what it is about horses—their teeth, their digestive system—they are pretty successful through it all."

In other words, once horses came through the evolutionary bottleneck represented by the scarce remains of *Epihippus*, they may have changed in some basic way.

"Horses are unique," he said. "Something happened to them that allowed them to adapt to whatever was thrown at them. Whatever it was that happened to them, it allowed them to track their changing environments so closely that we can use their fossilized teeth to track the history of climate change over the long term."

I had seen that modern wild horses were able to make small evolutionary changes that helped them adapt to where they lived, but I had no idea that this talent could be traced all the way back to a specific period in time, to a specific climate shift and to a specific tectonic event.

Forged in extreme warmth, then tempered by cold, horses managed to cross from the Eocene into the Oligocene as animals who had survived fire and ice and could withstand an almost unbelievable variety of conditions.

In another row of cabinets, Mihlbachler pulled out a drawer filled with ancient horse brains.

"These are Radinsky's brains," he said, with a certain reverence.

A leader in the field of paleontology, Leonard Radinsky died prematurely in 1985, saddening his colleagues so much that when one researcher found an early jaw of an animal in China that could have been a dawn horse ancestor, he named the fossil in the deceased scientist's honor: *Radinskya*.

What Mihlbachler found in the drawer were not really brains, of course, but horse brain endocasts—casts of the cranial cavity, the part of the skull that houses the brain, of extinct animals. Endocasts come in three basic varieties—those that are themselves fossils, having formed naturally of sand and other minerals inside an ancient animal's skull; those physically created by a research scientist who fills a fossil skull with composite material; and those that are virtual scans, made with modern computer technology. Some of the endocasts Mihlbachler found in the museum drawer were fossils Radinsky had worked with. Others were casts Radinsky had made.

Field researchers had known from early days about the existence of naturally formed endocasts. Workers had dutifully collected and cataloged them for future research along with fossilized bones, but not a lot happened with them research-wise for quite a while. The natural endocasts, often poorly formed, just didn't seem to yield much information. Most scientists believed there was little to be gained by studying them.

Then Tilly Edinger, a German Jew associated with the Senckenberg Museum before World War II, was fortunate enough to emigrate after the horror of Kristallnacht in 1938. In Europe, Edinger had already begun to show that endocasts could provide usable information because they revealed some details about the exterior shape of the brain. When she came to the United States, she, like Thomas Henry Huxley before her, found what was lacking in European museums: a consistent series of horse fossils. Realizing that she could follow the evolution of the brains of horses over 56 million years, she spent much of the following decade writing a major scientific paper. In her seminal 1948 "Evolution of the Horse Brain," Edinger asserted that even the evolution of the senses in horses could be traced, in a limited way, by studying the changes in endocasts. The result was the codification of an entirely new research field: paleoneurology.

Leonard Radinsky was one of her intellectual offspring. At the American Museum of Natural History, he followed up on Edinger's work and discovered that, interestingly, some of the details were wrong. Edinger had described the brain of the little dawn horse as being extremely primitive. Indeed, according to Radinsky, she had written that the early horse brain was "strikingly similar" to that of a marsupial mammal, the opossum. Radinsky found that Edinger had misidentified the brain she was studying. She was looking at the brain not of an early horse, but of a different animal.

In fact, Radinsky found that the brain of the dawn horse was "considerably advanced." Moreover, the dawn horse was showing some clues as to its future direction. The horse's olfactory bulbs, organs in the forebrain dedicated to the detection of smells, were already well developed. The neocortex, involved in what we call intelligence, was proportionally much smaller than it would be tens of millions of years later, but compared to other animals of the time, the horse had a "larger, advanced brain," the paleontologist Richard Hulbert told me. "This advanced brain could have conveyed competitive advantages over other Eocene herbivores such as more complex social behavior and better recognition and evasion of predators."

Radinsky compared the brain of the dawn horse with that of *Mesohippus.* He found that although this horse was only a bit larger than the dawn horse, he had changed in an important way. He had a substantially expanded frontal lobe. Among the new experiences *Mesohippus* would be able to enjoy because of this expanded brain, Radinsky wrote, was an enhanced sensitivity of the lips and the mouth.

So it seems that one result of the colder period that closed out the Eocene was that the lives of horses became sensually and intellectually enriched. The environmental challenge facing first *Mesohippus* and then *Miohippus* wrought a smarter horse with a greater ability to perceive and use information in the more open world in which he lived.

Some primates also seem to have started down a different evolutionary pathway following the arrival of the cold and the spread of grasslands. This is when, according to some researchers, some primate lines first evolved the three cones in the eye that allow us to see more colors than other mammals (including horses)—a change driven in part by the fact that as fruits became limited, primates needed to forage with greater focus to detect protein-rich young leaves, which are often red.

Our modern horse's ability to interface with the world via supersensitive lips, muzzle, mouth, and nose is an evolutionary gift resulting from the deep cold of the late Eocene, from *La Grande Coupure*, and from the transition into the harsh Oligocene.

"A cow could never play Mr. Ed," the paleontologist Donald Prothero once wrote, as a way of extolling the nimble lips of horses.

The path that led to Whisper's labial dexterity—his ability to use his lips the way I use my hands, to manipulate the latch on his stall door and pasture gate, to break into the grain bin, and to make a cup with his lower lip in order to drink out of my outdoor faucet—has some pretty impressive roots in deep time, and is associated with the sudden appearance of the Antarctic Circumpolar Current and the development of the Antarctic ice cap.

North America remained the center of horse evolution. *Mesohippus* eventually died out, leaving only *Miohippus*, the three-toed browser with slightly longer legs who roamed North America from about 32 million years ago to about 25 million years ago. Some descendants of *Miohippus* escaped to the Old World, where they evolved into strange forms like *Sinohippus*, a three-toed animal that looked something like a cross between a modern cow and a modern donkey. *Anchitherium*, a narrow-muzzled three-toed browser with a long neck that could reach up to eat the leaves of deciduous trees, evolved in North America and then spread to the Old World, including Africa.

Nevertheless, the real action in horse evolution continued to be in North America, where the Rocky Mountains continued to rise, where the continent's interior was drying, and where grass was spreading prolifically. A different kind of grass had evolved, one that could flourish in the most challenging of environments and could spread boldly into regions where the earlier grasses could not go. Horses, with their ability to run quickly over treeless grasslands and with their teeth that could, by now, grind up the toughest of fodder, also spread.

By the Miocene, horses filled so many different nooks and crannies that nobody knows how many species of horses roamed the world. We know of at least twenty different horse species in North America at this time. There were others in Asia, in the Middle East, in Europe, and even in Africa, where we primates were at that time also undergoing some pretty dramatic changes. There, during the Miocene, the first apes appeared, and then the first humanlike primates.

The diminutive dawn horses of the Eocene, so well-preserved in the Messel strata, had long since disappeared. But following the near-extinction crisis of *Epihippus*, horses once again became one of the world's dominant mammals. In fact, the Miocene, which extends from about 23 million years ago to a little more than 5 million years ago, was *the* epoch of the horse. The continued rising of mountain ranges all over the world—the Andes, the Himalayas, the Rocky Mountains—lifted the crust of Earth, changed global wind patterns, dried out continents.

"This was a major episode in the spread of grasslands," Mike Voorhies told me when I visited him. "You had a relatively new habitat and a family of mammals—horses—that could exploit that habitat in ways that other animals could not."

And as the interior of North America dried, and as grasslands, Candace Savage's "drought specialists," spread, horses adapted. At the beginning of the Miocene, horses with three toes were all there were. By the end of the Miocene, the three-toed horses, who had roamed Earth for more than 50 million years, had begun to disappear, although the process would take a while. Many of those had found their way from the center of North America all the way to Africa, via a route through Asia. Strangely, it would turn out that Africa, of all places, would be one of their final refuges.

4

THE TRIUMPH OF *HIPPARION*

> The history of the horse family is still one of the clearest and most convincing for showing that organisms really have evolved, for demonstrating that, so to speak, an onion can turn into a lily.
>
> —GEORGE GAYLORD SIMPSON, *Horses*

On the edge of northern Tanzania's grassy Serengeti, a tiny mare and her rambunctious foal hurried along a path that ran not far from an active volcano. The ominous cone, lurking high above the open plain, spouted irregular plumes of gas and ash much as a modern whale spouts water.

Most likely, the rumbling of the earth did not alarm the horses. It was about 3.6 million years ago, and where the dam and foal lived, light ashfalls were routine. The uneasy plain was dotted with such volcanoes.

The little mare, *Hipparion*, weighed much less than a modern horse, probably just a bit more than one hundred pounds. We would have recognized her as kin to the modern horse, but we wouldn't have mistaken her for one: She had three toes on each of her four feet. But by now, more than 50 million years after the dawn horses first appeared, the two side toes were very small, so small that nineteenth-century paleontologists, finding *Hipparion* fossils, thought the "extra" toes were "useless" vestiges left over from evolutionarily older species.

As she traveled, prudence was the order of the day. She had to pick her way lightly. The slippery surface under her hoofs was covered with loose ash—about as easy to negotiate as ice-covered ground topped by a thin layer of new snow. No modern horse would voluntarily choose to trot over such a complicated surface and risk a torn ligament or broken leg. Neither did this ancient mare. Instead, she adopted a four-beat running walk—a stabilizing gait that allowed her to move quickly but also have three feet on the ground at any one time. She was being cautious.

Her foal, however, was not. Perhaps too inexperienced to recognize the danger of falling—or perhaps just behaving in the brash way that babies sometimes do—the youngster wove back and forth from one side of his dam to the other. His path was haphazard. The mare seems to have felt no sense of urgency, no need to move with anything but deliberate speed. The same cannot be said for her baby.

At one point, the clueless foal ran right in front of the mare. The mare responded by glissading for an instant. Next, she steadied herself with her side toes. They may have been considerably smaller than those of her ancestors, but those side toes were still useful. Planted in the ash, they created a firm tripod that helped her stay upright. Having regained her footing, she traveled on.

Around the same time, a small band of our own ancestral relatives, with brains about one-third the size of ours, ambled along a pathway that, for a few footsteps, almost paralleled the horses' tracks. Standing up straight, with only two feet to worry about, these early hominids seem to have had an easier time navigating the terrain. Their footprints show no evidence of slipping and falling.

Two of these beings, members of a species we call *Australopithecus afarensis*, appear to have been walking side by side. They left footprints resembling those we make in beach sand. Their toes were pointed fairly straight ahead and there even seems to have been a bit of an arch that may have encouraged a rocking motion from heel to toe. The footprints of one of these side-by-side hominids measure almost twice that of the other set of footprints, so that we might in the twenty-first century wonder wistfully: Were they parent and child? Was the parent trying to lead the child to safety? It's possible.

But perhaps they were not walking together at all. One may have been following the other. Maybe it's only a coincidence that the foot-prints left in the ash made them appear to be companions. The precise details elude us. We can only see so much through the thick veils of time. We do know, though, that if all the *A. afarensis* prints on the scene were left at the same time, the pair was not alone: following them was at least one more two-legged walker. There is at least one more set of footprints in the ash.

What were they doing there? Were they out hunting? Again, it's possible. By this time, the tiny proto-thumb found in the Messel primates had developed into an opposable digit something like our modern thumb, so that these two-legged creatures would easily have been able to grasp and manipulate a naturally sharp stone as a tool, just as modern chimpanzees can. Or perhaps they weren't "hunting" in the modern sense of the word, but were instead out "power-scavenging"—forcing predators like hyenas to give up the prey they had just taken down.

After these horses and hominids passed by, more ash fell. The heat of the African sun baked the prints into the ash and then, over the ages, soil and other detritus covered the evidence. A natural time capsule, this fossilized record of horses and hominids lay buried for several million years. Then the wind and the rain washed away the protective layers.

Slowly, the prints began to reappear: proto-human and proto-horse, striding along under a sky full of volcanic ash. This behavior was not evidence of any earth-shattering event. There's nothing in the tracks showing that the animals were in a mad dash. It's not like Pompeii. Indeed, the day seems to have been rather mundane. There were no disasters in the offing, no mass extinctions just around the corner. It was rather quotidian. Just a routine scene under the African sun. A moment caught in time. Like family photos at the beach.

Nevertheless, the tracks, despite their mundanity, are compelling. We can see here, in these fossilized prints, our own continuity and our own futures and the coming era when horse and primate, finally reunited after their expulsion from the Garden of Eden Eocene, would partner up to depend on each other for food and survival. The day of *H. sapiens* was nearing. And so was the day of the one-toed horse. Indeed, on the other side of the Atlantic on the North American plain, the modern horse *Equus* had already arrived.

On that day 3.6 million years ago, did the proto-horses and proto-humans encounter each other? I'm betting against it. After nearly tripping up the mare, the foal traveled on heedlessly. We can see the baby's fossilized hoofprints next crossing the fossilized footprints of the *A. afarensis* group. The foal's tracks show no hesitation during that crossing. Then, as today, it's unlikely that a callow baby would have blithely met up with these proto-people without at least one surprised side step. The foal's tracks show no such startle response. We have to remain content that these two species shared the same plain, but that, at that instant at least, they probably missed each other like the proverbial two ships in the night.

Hipparion and *A. afarensis* were not alone in this landscape. The site is covered with at least sixteen thousand fossilized tracks of ancient mammals and other animals, all probably made within days or weeks of each other. Scientists have uncovered tracks of large cats, giraffes, elephants, hyenas, ostriches, and so much more. There's even a fossilized insect trail.

The story of the tracks is a superb tale of how serendipity and science sometimes work together. Local people had known there were ancient tracks in the area, but no one had really "noticed" them in a significant way. They didn't fully grasp how much information the tracks held. We routinely see amazing things in the world but too often have tunnel vision and don't pay attention to what we're seeing, and so it was with these tracks. No one had taken the time to stop and think about the volumes of ancient behavior recorded there.

It took an accident and a lightbulb flash of insight from a young scientist for the phenomenon to win the attention it deserved. In 1976, a thirty-year-old researcher, Andrew Hill, now an anthropologist at Yale University, drove down from Nairobi with friends to visit the site, near the present-day village of Laetoli, at the invitation of Mary Leakey, a well-known scientist. Leakey was focused on finding fossilized hominid bones and skulls. She was not thinking about animal trackways.

Neither was Hill. But one day Hill and some buddies, engaged in their own particular form of horseplay, began throwing elephant dung at one another. Hill slipped and fell, he told me. At ground level, he noticed a few of the fossilized tracks from a new angle.

In a gratifying eureka moment, he associated those tracks on that African plain with an illustration of fossilized raindrops he'd seen in a book about geology by the influential nineteenth-century scientist Charles Lyell. In that book, Lyell discussed the scientific value of these kinds of records: Something as simple as fossilized impressions of raindrops, Lyell wrote, can—if paid attention to—tell us a great deal about what life was like in prehistoric times. For example, Lyell wrote, take note not just of the raindrops themselves, but of the size of the raindrops.

Lying there on the hot African plain, Hill remembered Lyell's admonition and realized that the animal trackways he was looking at that day could well contain a whole library of information. When researchers began a full-scale investigation of the footprints, they found that many, many animals had passed through the same region during that time, and that many of them had left evidence of their presence and their behavior in the ash.

The finding of footprints left by ancient members of the human family made headlines worldwide. It settled (to the degree that anything is ever settled in paleoanthropology) a long and sometimes rancorous debate: Were our ancestral relatives beings who walked on all four limbs? Or were they able to stand upright? How early did bipedalism appear? Laetoli shows that *A. afarensis*, thought by many researchers to be intermediate between apes and modern humans, not only walked upright—but had a heel, a big toe, and even an arch.

By 1987, a two-volume set of papers appeared that included a pile of maps with the trackways of all the animals found to date. I unfolded the maps on my dining room table one afternoon, piecing them, jigsaw-puzzle-like, together, until they covered the table. You could see almost the whole of the animal world that then inhabited this region of Africa—a cornucopia.

For anyone who loves the detail you find in maps, these sheets of paper were dazzling. There were the footprints of the ancient hominids right in front of me. I could imagine the three-toed horses, wandering under the African sun, steadfastly enduring that light rain of volcanic ash. Looking at the maps, I felt in my dining room the way I'd felt at Polecat Bench and at Messel: as though I'd crossed some sort of fourth-dimension time barrier and entered forbidden territory.

Seeing the footprints right there on my table of so many animals who had walked the Serengeti millions of years ago was a metaphysical experience, but the tracks, it would turn out, also revealed important information.

Eadweard Muybridge's sequence photographs of the horse's gallop and walk

For example, in that same volume I found an article addressing an issue that has long bothered horse aficionados worldwide: Do horses come by the four-beat single-foot gait naturally? Or is special breeding necessary? Horses today have three basic gaits: the walk, the trot (or pace, for some horses), and the canter, or lope. (The gallop is a faster version of the canter.)

But some horses can also do a four-beat gait that's faster than the walk and is sometimes called the single-foot or running walk. Because it's both comfortable and fast, riders treasure this gait. Since horses tire less using this gait than using a canter or gallop, it allows horse and rider to cover a lot of ground in a day without the horse becoming overly tired. But only certain breeds of horses, like the Tennessee Walking Horse or the Icelandic horse, do this naturally. Do they have this ability because it was bred into them, or did at least some horses always have this ability?

The Dutch scientist and avid horse enthusiast Elise Renders decided to use the Laetoli tracks to try to answer that question. When she first read about the *Hipparion* tracks, she, like me, was fascinated. A record of the specific behavior of two individual, now long-extinct horses made her wonder if she could discover how these horses actually moved. She also wanted to know if the two tiny side toes on each foot were really superfluous. Or did they have a function?

First she thought about the gait the *Hipparion*s were using. She couldn't tell just by looking, so she made a cast of the original Laetoli tracks. Then she made similar copies of modern horse tracks. She had these horses walk, trot, and canter. She compared the modern tracks with the ancient tracks and was surprised to find that none of the modern tracks matched the *Hipparion* tracks. Apparently the mare and foal were not using the walk, trot, or canter.

Then she began looking at modern horses using the running walk, a four-beat gait. She hit pay dirt. She measured the distance between the front foot and the hind foot, and then measured how far the hind foot stepped over the front foot. She found that both the little *Hipparion* mare and her foal were moving in a four-beat gait.

"This mare and foal had a natural traveling gait," she told me.

In a regular walk, a horse moves one front foot, then the opposite hind foot, then the other front foot, and then the opposite hind foot. In the running walk, the order is different. The horse moves first one front foot, then the hind foot on the same side, then the other front foot and then the hind foot on that side. The gait is also sometimes called the "broken pace."

The *Hipparion* had adopted this gait, which allowed the mare to move faster than a normal walk but still have three feet on the ground. It's a very secure gait, one that allows the horse to move faster but also helps the horse not slip.

"Everybody in the horse world today is familiar with the walk, trot, and gallop," she told me, explaining why she looked at those gaits first. "We have the belief that the running walk is an artificial gait or a show gait," she explained, adding that when she realized that was what she was seeing in the *Hipparion* footprints, "it just seemed to be the most natural thing, especially when you take the slippery soil into account. A horse that can do a running walk won't trot when the soil is not good enough. With my horse, as soon as the soil is uneven or slippery or as soon as she has some kind of stress, she will go into a running walk."

The ability of the horse to use a running walk has long been known to be a heritable trait, but it was thought to be a trait that had been *bred into* domesticated horses. Until Renders's research, few people thought the gait was natural not just to a few breeds of modern horses, but to at least one mare and foal 3.6 million years ago.

Renders also found that the *Hipparion* mare had an added measure of security to keep her from slipping—those tiny side toes. At first glance, the side toes don't seem to make sense: we assume that horses "should" have one toe and that horses "improved" when they got rid of their extra digits. We tend to look at the "extra" toes of earlier horses as vestiges, excess clay that the sculptor evolution needed to remove.

But Renders discovered that horses actually *used* those side toes. They weren't simply relics from a bygone time. In the case of the Laetoli mare, those toes served her well on that particular day. A modern horse traveling over that terrain might have fallen and broken a leg (this happens to many modern horses trying to cross ice, for example), whereas the *Hipparion* mare didn't break her stride. Instead, the evidence shows that when the mare slipped, she planted her side toes on the surface. The three toes—one main hoof and two supporting toes—made a triangle in the ash and kept her upright. You could think of them as similar to training wheels on a child's bicycle.

Because of the toes and because of this horse's natural four-beat gait, and for many other reasons, *Hipparion* horses were enormously successful. They first appeared in North America roughly 17 million years ago and quickly multiplied into countless species. Eventually, they spread all the way to Africa.

The several earliest *Hipparion* species were larger than dawn horses, but much smaller than the mare and foal traipsing the Laetoli plain. But while the dawn horses were evolutionarily conservative, *Hipparion*s excelled at evolutionary experimentation. This meant that as the natural world changed gradually through the Miocene epoch there always seemed to be at least one new type of *Hipparion* able to take advantage of new opportunities. Some *Hipparion*s evolved into larger animals with larger teeth. Some stayed small. One group even evolved teeth that would *literally* grow continuously until the age of about five.

The large *Hipparion* group, with many different genera, had a talent for adapting to whatever natural system they encountered. Because of this malleability, they spread across the world. Their fossils have been found in the soggy areas of Florida, a state with so many fossilized horse bones that you may well find one as you walk the Gulf Coast beaches. They turn up in the dry areas of Asia, in the Middle East, in Greece, on several Mediterranean islands, in Europe, and even north of the Arctic Circle on Ellesmere Island.

*Hipparion*s must have been able to eat an astonishing range of forage. With a light body, complicated teeth, a deeply grooved astragalus, and a useful three-toed foot, *Hipparion* was "one of the great animal travelers," wrote George Gaylord Simpson, the mid-twentieth-century expert on horse evolution who was based at the American Museum of Natural History. As with the dawn horses, the first appearance of *Hipparion* in mid-Miocene North America correlates closely with a time of warmer and wetter weather. It was not as hot as the Eocene, but the global climate was warm enough that the ice in Antarctica melted a bit, allowing a variety of plants, including small trees, to grow even on the edges of that continent.

We know this because Sarah Feakins, a molecular stratigrapher, has studied leaf waxes of plants growing in Antarctica from this period. She found that just when horse populations were exploding in North America, the temperature in Antarctica was about 11 degrees Celsius above what it is today.

Curious about how 20-million-year-old leaf waxes could provide such information, I called her. "What," I asked, "do leaf waxes have to do with global temperatures?"

"Look at the waxy coatings on your house plants," she said. "You're seeing very resilient hydrocarbons, evolved to protect the leaf. They're very resilient molecules. When the plant dies, these molecules don't get eaten [by bacteria]. These waxes wind up in the ocean."

Which is where they stay, in marine sediments, until researchers come along and take deep-time seabed cores that Feakins and others can open up and study. By looking at the isotopes of carbon atoms and calibrating them to tables that scientists have spent decades working out, they can determine environmental conditions. Feakins found that in the throes of a general worldwide warming trend, even Antarctica had become less icy. Scientists suspect that these relaxed temperatures in the Antarctic began a cascade of events, from shifting marine currents to changes in rainfall, that accelerated this period of global warmth. Of course, we don't know exactly which came first—the warmer Antarctica or the warmer planet itself. But we do know that during this period of 5 million years, a multitude of horse species evolved in North America.

And yet—there were, at this time, no horse species living in Europe. Relatives of horses like tapirs and rhinos were common—but not horses. The absence of horses in Europe and Asia ended about 12 million years ago. One single *Hipparion* species escaped from North America to colonize Asia by traveling over Beringia, the swath of land connecting Alaska and Siberia. In the Old World, this *Hipparion* struck it rich. The Asian steppes were full of grazing opportunities.

And horses, with their ability to eat so many different kinds of food, took full advantage of the forage around them. One scientist wrote enthusiastically about *Hipparion* "galloping" from Alaska to Spain.

Which brings us back to further discussion of Charles Darwin's stress headache. The sudden and prolific appearance of little *Hipparion* in Europe was one of his life's major conundrums. To Darwin, the little horses with the two tiny side toes seemed to materialize out of thin air in the Old World. And yet, Darwin, as we said, was not comfortable with this kind of change. Although life clearly evolved—such change should not stand out in such an apparently gross manner.

Hipparion made evolution seem like a magic trick. Remember: while Darwin was struggling with the question, scientists still believed that horse evolution occurred not in the New World, but in the Old, so that the "sudden" evolution of a tremendously prolific type of horse seemed almost like a kind of alchemy.

One of Darwin's British critics had accused him of claiming that life forms appeared on Earth in "higgledy-piggledy" fashion. This bothered Darwin a great deal, because he was anything but anarchic—and because *Hipparion's* sudden appearance in Europe did, at least at first glance, *seem to be* higgledy-piggledy. Until Huxley visited O. C. Marsh at Yale, Darwin couldn't reconcile the sudden appearance of *Hipparion* with his understanding of how life changes over time: "That many species have been evolved in an extremely gradual manner, there can hardly be a doubt," he asserted in *The Origin of Species.*

Hipparion wasn't the only animal that worried Darwin. The horses came along with a whole host of other animals that seemed to scientists to have flooded Europe, almost as though there was a second *Grande Coupure.* In fact, there was indeed a boundary line in the rock record, just as there was in the 34-million-year-old event.

This clear line in the rock was so obvious that nineteenth-century scientists could easily recognize it and, since the newly arrived *Hipparion* were so plentiful at the time, the event was called the *Hipparion* Datum. (For paleontologists, this is akin to the iridium layer, the result of the asteroid impact, that demarcates the end of the Age of Dinosaurs and the beginning of the Age of Mammals.) The *Hipparion* Datum helps paleontologists working in Asia and Europe understand the age of the rocks they are looking at: if *Hipparion* bones are present, the rocks must be less than 12 million years old.

But *why* did *Hipparion* appear in the rock record as if these three-toed horses had traveled almost instantaneously from the farthest reaches of Asia all the way to Spain? The answer to that mystery has only recently come to light: the horse was following the grass. The triumph of *Hipparion* was also the triumph of grass.

We underestimate grass. The ten thousand or so varieties of grasses that cover Earth today take up an estimated 30 percent of our planet's land surface. We see grass all around us, but what we see aboveground is not where the action is—not the part of the plant that conquered the world. About 80 percent of a grass plant—the most important part of the grass plant—lives *under* the soil surface, in roots that can intertwine in networks so thick that early European settlers sometimes needed teams of twenty or even thirty horses to plow the prairie sod for the first time.

Grasses store away so much carbon in their massive underground root systems that the paleontologist Gregory Retallack and many others suggest that grasses ultimately became as important an evolutionary force as tectonics. When he told me that, I was skeptical. But the more I read, the more I realized that many people agreed with him. My skepticism was probably no more than a typical case of familiarity breeding contempt: having mown a suburban lawn, I thought I knew what grass was. I was wrong. Very wrong.

Grasses are too modest for their own good. They don't get respect. They're often easily trod upon or weeded out of our gardens. Because they seem simple, we assume that they appeared early in evolution. Nothing could be further from the truth. "In fact," wrote Candace Savage in *Prairie*, "they are highly evolved organisms, especially adapted to cope with extreme climatic uncertainties, including frequent drought."

For example, the aboveground blades of grasses can sometimes curl themselves up during droughts to conserve as much water as possible. In other words, when trees die out—grasses live on.

Of course, most of the grasslands we see today are just pitiful remnants of what was here prior to the expansion of farming. Before that happened, a rider on horseback could travel through some expanses of tallgrass prairie and not see over the tips of the tallest grasses. I had read about this, but, to be honest, didn't quite believe the authors. I suspected them of hyperbole. So I decided to see for myself. This isn't as easy as it sounds, as almost all the original tallgrass prairies are gone. But in Illinois, of all places, at the Midewin National Tallgrass Prairie, just a bit southwest of Chicago, scientists and volunteers have worked hard to rehabilitate some of the land. Some of the stems of big bluestem, a species of grass, rose to nearly ten feet.

These are not your city park grasses. To ride through a field of big bluestem must have been like riding through a forest, but one perhaps much more threatening than a forest of trees, since you wouldn't have been able to see beyond the next few blades. We think of prairies as being places of wide vistas, but if the tall grasses were healthy, visibility at times might have been almost nil. Anything could be hiding in those grasses, only a few feet away, and you might not know until it was too late. A healthy grassland can create its own kind of junglelike environment. No wonder horses are so easily startled. Of course, when I was in Illinois, I didn't have to worry about saber-toothed cats or packs of dire wolves creeping up through the grasses behind me, but I did get a whole new understanding of why horses are so cautious and always listening, alert for even the smallest rustling.

Both bison and horses love to graze on big bluestem, but it took quite a while for the grass to evolve. Grasses, of course, are a type of flowering plant, so they had to wait for the Cretaceous Terrestrial Revolution. When they first appeared (no one knows exactly when that was), they were simple and slow on the uptake. They certainly did not conquer the world by storm. While the world was warm and wet, grasses could grow along the edges of the forests and in glades and clearings. But when the Eocene ended and the world became drier and the trees slowly died back, the grasses—plants that could protect themselves from drying out by growing deep, thick root systems—began to spread.

They were still somewhat handicapped, though, in that they did not do well in some conditions, like strong sunlight. Before they were truly able to take over the world, grasses needed to come up with an evolutionary innovation. It was after that innovation that *Hipparion* began spreading worldwide. A new type of grass evolved that could endure some truly harsh conditions, such as very high temperatures and drought.

This innovation changed the world. The marriage of these two types of grass—cool-season grass and warm-season grass—created a power couple that the whole planet would embrace. If you think about this, it's pretty obvious: cool-season grasses grow better during the cool season, and warm-season grasses grow better during the warm season. When you get down to individual species, it's more complicated than this, but for our purposes, all we need to know is that neither type of grass is "better" than the other. Each has its own advantages and peculiarities, but together—what a marriage they make.

Sometimes the two grasses grow in one area, and when one type dies back, the other thrives. You can see this phenomenon by looking at a typical suburban lawn. Many have at least small amounts of both types of grasses. Some green up in the spring, then become brown in the summer, and then green up again in the fall. Some green up during the summer. If you see patches of brown in a green summer lawn, you may well think that the brown grasses are dead, but they're not. They're just resting until cooler weather returns.

In wild lands, something similar occurs. Cool-season and warm-season grasses may both grow in the same general region, so that horses could, for the first time, count on having plenty of green food for much of the year. When one type of grass was brown, often the other type was green. All the horses had to do was know where to find which type of grass during certain seasons.

Additionally, the proportion of the two types of grass in an area could change over time. Weather conditions during one decade might favor cool-season grasses. But the next decade the weather might change to favor warm-season grasses. This made a grassland *system* that was much hardier in the long run.

The initial spread of the two types of grasses correlates closely in Africa with the triumph of *Hipparion*. Sarah Feakins has also studied cores of deep-sea sediments from the Gulf of Aden, just off the East African coast. Her research shows that in East Africa, just after the 12-million-year-old *Hipparion* Datum, both kinds of grasses ebbed and flowed over the landscape in response to shifting patterns of rainfall and temperature.

Feakins's colleague Kevin Uno has studied the ways in which some African animals changed in response to the new grasslands by examining the teeth of horses, rhinos, and other species from about 10 million years ago. After studying the wear patterns on the fossil teeth, Uno found that horses were uniquely flexible in their eating habits. Within less than a half million years after the new grasses—called C_4 grasses by scientists—first appeared in Africa, horses were eating them. And the horses who were eating these grasses turned out to have different teeth.

To us, of course, a half million years seems like a long time.

"That's a startlingly fast shift over geological time scales," Uno told me.

He studied the same type of tooth from the same horse species and found that before the C_4 grasses spread, *Hipparion*s had tooth surfaces that contained sharp, knifelike serrations, indicating that the animals were eating food that didn't need much grinding. About a half million years later, when the grasslands spread, the *Hipparion* teeth had flatter surfaces. The horses living in the same area of Africa had to grind their food a whole lot more before swallowing. Their teeth showed the results.

Horses were once again exhibiting their exceptional flexibility. Uno found no other animal that showed this kind of quick adaptation.

"There are two strategies when the environment changes," paleontologist Richard Hulbert once told me. "You can find the few places left in the world that don't change. Or, you can adapt."

Horses adapted.

Even as *Hipparion*s spread through the Old World, they were already being replaced in North America, where the spread of the new grasses encouraged the evolution of a larger, faster horse. We know something about this transition because of yet another volcanic eruption, but one that was much more violent than the one that left the tracks at Laetoli.

On the day of the North American disaster, which occurred about 12 million years ago, horses of several different species were grazing a grass-covered plain in what's now Nebraska. Perhaps a few of the horses sheltered themselves from the sun under the walnut and hackberry trees that then dotted the landscape. Perhaps a few nibbled at leaves on shrubs. But most were probably busy eating the grasses that had replaced the Eocene wetlands. Accompanying the horses were hump-free camels, saber-toothed deer, strange rhinoceroses, several species of dogs, elegant cranes, and long-tailed secretary birds.

While the horses grazed, a thousand miles to the northwest a supervolcano exploded. Unlike the Laetoli volcano, the Bruneau-Jarbidge eruption was deadly. Its ash spread across hundreds of thousands of square miles, including the plain where the horses grazed. Tiny bubbles of silica—like soap bubbles but much, much smaller—emerged from the volcano and then shattered, creating a multitude of glassy, curved microscopic shards that wafted, like parachutes, a thousand miles distant on the winds that blew to the east. When they finally landed on the grass, the horses and other grazers breathed these shards into their lungs while they ate. Imagine taking several glass Christmas tree ornaments and pulverizing them with a hammer, then spreading those sharp-edged infinitesimals out over a field of grass. That's what the animals inadvertently ingested.

Paleontologists have worked out the order in which the animals died. First, tiny birds fell out of the sky. Their lungs were the smallest and the most easily damaged. Then the smaller land animals succumbed. Then slowly, the larger animals, including the horses, died, their lungs destroyed by the glass-like micron-size silica that entered with every inhalation. The last to die were the largest animals with the largest lungs, the rhinos.

It was a slow, agonizing death. It must have hurt the horses to breathe, but of course they had no choice. In their misery, many of the animals sought out a local water hole, no more than a slight indentation in the plain that filled when the rains came. There wouldn't have been a lot of water, but apparently even this tiny oasis offered some kind of solace. Perhaps the horses wanted to drink, or perhaps they were just seeking wet mud to cool their terrible fevers. Their slow suffocation caused bone decay, lung damage, swelling of body organs—damage that's still visible even today on the many skeletons that have been left in situ for visitors to see.

After they died, the prairie wind continued to blow. The light ash covering the plain drifted over the cadavers, entombing them.

Twelve million years later, in 1971, the paleontologist Mike Voorhies and the geologist Jane Voorhies walked through the area and found an American *Lagerstätte*, one of those sites like Messel where, no matter how hard you work, there will always be more science to be done. Like Messel, Nebraska's Ashfall Fossil Beds were unique. Messel fossils were flattened between paper-thin layers of clay and algae. At Ashfall, where the falling ash was light, the animals were preserved in three dimensions, similar to Vesuvius victims. The same tiny glass shards that killed the animals also preserved them. Voorhies likens the preservation material to the ultralight packing peanuts used for mailing delicate objects.

"The reason those peanuts work so well is that they're curved," he told me. "If the peanuts were flat, they'd settle out. The thing about the curved shape is that it traps air. When the ash came out of the volcano, it came out as bubbles, and the bubbles crashed against each other and made little curved pieces of volcanic glass."

You can't see any of these curved pieces with the naked eye, of course. When I visited the site, I rubbed the silica bits between my thumb and finger and it felt powdery, like flour. It seemed innocent enough. But then I felt a few mild, almost imperceptible pinpricks in my skin. I realized that if I breathed in this material, in only a short period of time, a few days perhaps, it could destroy my lungs.

Ashfall, a National Natural Landmark open to visitors, marks a key turning point in the history of the horse. That its preservation is so spectacular is like a gift from the gods. Even the most casual visitor, like me, can walk into the building where the skeletons have been left in the ground and see a flash point in time when horses (and other animals) lived on the cusp of global change. Paleontologists have found *five* different intact species of horses at Ashfall, ranging from three-toed horses who were only a bit bigger than the dawn horses to one-toed horses who were almost as large as our own modern horses.

What's interesting about this, Voorhies told me, is that only a few million years earlier, at least twenty species of horses roamed this region at the same time. Then a drying event occurred. Geologists know this because they've found a layer of caliche—hardpan—just below the layer containing the Ashfall horses. This hardpan is evidence, Voorhies said, of a "significant drying event" that may have helped whittle down the number of horse species in the area from twenty to five.

Most of the time when fossils are found, they're found in bits and pieces. A piece of the skull in one place. A bit of tail bone ten feet away. Perhaps scavengers pulled them apart, or maybe the animals died in floods and the flowing water scattered their bones. It's not always easy for researchers to figure out how the bones fit together, but at Ashfall there's no uncertainty.

The skeletons are intact—thanks in large part to yet another bit of serendipity. Millions of years after the horses died, an ice sheet grew over much of the northern half of the continent, shredding everything in its path. Because of the destructive nature of the ice, we lack good records of the life forms it covered.

But the ice sheet stopped just seven miles away from where the horses lay fully articulated and covered in ash. Had it continued, there would have been little, if anything, for Voorhies to find.

I asked why the ice sheet hadn't come farther south, but was quickly corrected on my wording.

It wasn't that the ice hadn't come farther south, Voorhies explained, but that the ice couldn't flow that far *uphill.* The most recent ice sheet to cover much of North America was, in some places, miles thick. The weight of all that ice pushed lobes of ice forward in pulses, but the tips of those pulses could only climb so far above sea level before they ran out of power. In eastern Nebraska, the lobe pushed its way all the way up to 1,650 feet above sea level.

Ashfall, Voorhies said, is 1,700 feet above sea level.

When I thought about that, I thought about all the paleontological evidence of horses that must have been destroyed by the relentless ice.

How much more we would know, I said to Voorhies, had it not been for the ice.

"There's so much that goes on in the biological record," he answered, "that doesn't leave any trace. Fossils are just a pitiful remnant." His voice was full of regret.

Because the Ashfall skeletons *are* intact—and because, remarkably, some of the flesh was desiccated rather than destroyed by bacteria—there's a wealth of information available. Researchers can see the structure of the cartilage in the horses' legs, the connections of the horses' bones with tendons and ligaments, and even what plants the horses ate. Marvelously, the neck flesh on one animal was preserved. We know that the volcano probably exploded in late winter or early spring, because foals have been found in some of the mares' uteruses, and foals, Jason Ransom and other colleagues have found, tend to be born in the wild in late winter or early spring.

The Ashfall bones will remain where they have rested over the past 12 million years, available during the summer months for the public to see. Excavators are carefully brushing the ash from the fossils, but the fossils themselves will stay in place rather than be removed. Voorhies prefers this approach.

I asked him why.

"I felt like I was vandalizing the site," he told me. "There's so much more information here, with the fossils in situ." For example, by examining the fossils in situ we can recognize that the multitude of animals found here are not touching each other. Had the animals died in a sudden panic, they ought to have been grouped together in terror. But instead, they are all separate.

It was as though, Voorhies said, touching each other would have been too painful.

Ashfall is a snapshot of a critical moment in horse evolution. The five species of horses preserved in this ash include three *Hipparion* species. In all three, the side toes are much more pronounced than the side toes of the much younger Laetoli *Hipparion*. These side toes were definitely functional. In the mud surrounding the water hole the *Hipparion* left tracks, as at Laetoli.

"You can see what you'd expect to see," Voorhies told me. "They look like tracks left today by the unshod hoof of a modern horse. But then, behind the main hoof are the impressions of the side toes. Looking at the anatomy of the foot, you can see that the side toes had their own sets of ligaments."

But, phenomenally, at Ashfall there are also *one-toed* horses. In other words, at Ashfall, we can see the precise point in time when horses evolved the modern hoof: horses with three toes and horses with one toe lived simultaneously. So much for the Victorian ideal of evolution. Here, at Ashfall, is a clear overlap showing that evolution is not linear—not one thing, then the next thing, but many different things, sometimes all at the same time.

Even more surprising is the fact that some of these one-toed horses, called *Pliohippus*, lay at the site beside *Pliohippus* horses with three toes. In *Pliohippus*, however, these side toes were useless.

I asked Voorhies how he knew they were useless.

"If you look at the bones in the side toes [of these *Pliohippus*], you can see that they did not have ligaments. These side toes would basically be functionless. In not too many generations," Voorhies said, "they would have disappeared completely."

"This is in one genus?" I asked.

"Just one *species*," he answered.

I was amazed. One species with two clearly different styles of feet. Were the differences irrelevant, like black hair versus brown hair? Or did the difference between three toes and one toe dictate important differences in the lifestyles of the individual animals? Were both types living equally successfully? If so, why did the three-toed horses disappear entirely?

Perhaps Darwin would have been surprised to find one species with two different kinds of feet, but he might also have been pleased. What better proof could he have asked for that his theory of evolution was correct? Twelve million years ago, 8 million years *before* the *Hipparion* mare and foal crossed paths with *A. afarensis* at Laetoli, in the Northern Hemisphere all the way on the other side of the world, a one-toed horse was slowly evolving and also, apparently, slowly overtaking horses with three toes.

Standing there at that site, seeing right before my eyes clear evidence that horses changed in response to a changing planet, was exhilarating and a little bit eerie. I tried to imagine what Darwin would have felt if he had known about this site, or even been able to visit it. The visit would have confirmed his theory, but it might also have refined it a bit. Perhaps he would have understood that evolution was not about "improvement" or "direction," but just about fitting in. In some situations, right there on the North American plain at that specific time, three-toed horses held an advantage. But in other situations, one-toed horses excelled.

So why did horses ultimately evolve to have only one toe? Traditionally, paleontologists have explained this phenomenon in terms of an improved ability to run over open areas in order to escape predators. The paleontologist and horse expert Christine Janis has suggested that the one hoof may also have allowed horses to roam farther in search of food. Three toes worked well in an Eocene world. In a drier world of grass-covered plains that provided mostly firm footing, running on a single toe had clear advantages.

I asked Voorhies if it would ever be possible for horses to re-evolve three useful toes—say, if the world returned to Eocene conditions.

Probably not, he said.

This is worth thinking about. When horses first appeared 56 million years ago, they had a wide array of options available, evolutionarily speaking. But once an animal begins to follow a certain path—running on one toe, for example—he may become so specialized that he can never go back. His fate is, to some extent, determined.

Voorhies agrees that many of the adaptations of horses during that time had to do with the proliferation of grass. By the time the Ashfall event occurred, grasses had become quite clever. They had learned to spread themselves by piggybacking onto animals. Some of the seeds of various grasses had tiny "hooks" on their seed coats. These hooks could attach themselves to the coat of whatever animal was passing by and get the seeds a free ride to some place where they could put down roots.

"If you walk through a field of high grass and get things stuck to your socks, you'll know what I'm talking about," he explained. "There's a type of grass called needle-and-thread grass, where the seeds are protected by silica." This grass has a silica-covered "needle" that can prick your finger as easily as a sewing needle and is excellent at embedding into an animal's hide, or, in our case, into our clothing. It takes days to get all of them out of your socks and off your shoes and pant legs. In this way we, too, are unwitting actors in the plants-versus-animals knockdown.

At Ashfall, however, the triumph of one-toed horses was not yet in full swing. The paleontologist Darrin Pagnac and his colleague Nick Famoso studied a large number of horse fossils and found that 78 percent of the horses had three toes, while only 22 percent had one. Judging by these proportions, the future of three-toed horses seemed secure.

However, the world continued to change.

Ultimately, only one-toed horses survived. As evidence, paleontologists point to Idaho's Hagerman Horse Quarry, where a number of broken-up horse bones have been found. Researchers assembled the fossils and determined that there were about 200 individual horses and that, a little more than 3 million years ago, they had all died, perhaps in a flash flood, then washed downstream until they came to rest along the riverbank.

All the horses at this site had only one toe. And they were all members of the same species—*Equus simplicidens*, the foundation animal for all modern horses, including our modern riding horses, zebras, asses, and the Przewalski's horse. In a sense, *Equus simplicidens* is akin to *Epihippus*, in that the future of horses depended on just one species.

But unlike *Epihippus*, *Equus simplicidens* was not rare. He was prolific. His long legs, his single-toe hoof, his flexible digestive system, his large brain and his hardy teeth—all combined to create an animal well suited to North America at that time. His evolutionary history had made him capable of withstanding an extremely varied range of environmental stresses and of eating all kinds of foods that other grazing animals would pass up.

In short order, *E. simplicidens* spread, spawning many species.

Then something—some kind of perfect storm—occurred. After 56 million years of evolution, the horse became extinct in the Western Hemisphere.

Why? If horses had managed to survive sudden temperature increases and decreases, and if they had managed to learn how to eat silica-covered grass when grapes disappeared, and if they gave up four toes on their front feet in favor of one—all in order to survive—then why were there no horses at all when European colonizers arrived in 1492?

Darwin looked for an answer to this riddle, but couldn't find one. Over the past 150 years, many of the other riddles of horse evolution that perplexed Darwin have been solved, but not this one. The problem remains the subject of heated, and sometimes overheated, debate. Indeed, these disagreements are sometimes representative of "science at its ugliest," to borrow a phrase from the biologist Bill Streever.

I wanted to know two things: What happened to the horses of the Western Hemisphere? And—how did the science become so controversial?

5

EQUUS

A hoof is like a second heart in a horse.

—J. EDWARD CHAMBERLIN, *Horse*

The last meal of the golden-coated Yukon horse was buttercups.

It took 100 million years for the Cretaceous Terrestrial Revolution to get those buttercups up onto that Ice Age northern plain. And it took about half that time for the scampering, warmth-loving dawn horses of the Eocene to transform into *Equus*, an animal capable of surviving in the frigid Arctic in order to eat those flowers.

But finally, after tens of millions of years, in the Yukon horse, the mosaic of the modern horse—the high withers, the hocks well off the ground, the spine and legs specialized for speed and endurance, the single-toed hoof, the sensitive muzzle, and the deep-jawed mouth with huge, durable teeth—was present in one powerful package.

In 1993, the miners Lee Olynyk and Ron Toews were digging for gold near Canada's Dawson City, just a bit south of the 60th parallel, close to the Arctic Circle. One September day, as operations were winding down for the year, their machines took a chunk out of the black muck oozing from the bank of Last Chance Creek.

"What's that?" Olynyk's young son asked.

Olynyk took a look. It was the carcass of a horse. Not just the bones, as at Ashfall, or an imprint, as at Messel, but a carcass of flesh and hair and tendons and mane and tail and gut and intestine. As well-preserved as if you'd just pulled it out of your freezer. But the animal seemed a bit odd. It wasn't a horse as much as a pony, Olynyk decided. He figured maybe it was one of the pit ponies the old miners used to pull carts in the underground mines.

"It smelled strong, fresh, like a horse smells when you just come in from a ride. A nice horsey smell," he told me. "The leg came out. Then we saw the hide."

The more he looked, the less he bought into the pit pony idea. It just didn't feel right. So he made some calls. Paleontologists arrived on the scene, pulled the horse out, and sent the flesh to a laboratory to find out its age.

The carcass turned out to be from the Ice Age and was almost thirty thousand years old. This horse lived in the Arctic Yukon at the time when, at the other end of the dry-adapted Eurasian steppes, Paleolithic artists were painting horses on the cave walls of France and Spain and craftsmen were decorating their weaponry with horse images. The Vogelherd horse had already been around for a few thousand years.

This particular Yukon stallion had a yellowish coat and a long, flowing blond mane and tail. This alone surprised the experts, since it was commonly believed that before domestication all horses had short, bristly manes and stumpy tails.

Olynyk's horse stood about four feet at the withers and had a disproportionately large skull (by today's standards). He was a tiny thing, weighing more than the Laetoli mare's hundred pounds but a lot less than the thousand pounds or so of a well-nourished modern horse. Quite stocky, with thick legs, a clunky head, and a cold-adapted convex nose, he would have been called a hammer-headed horse by the old American cowboys.

DNA testing shows that he is a thoroughly modern *Equus*.* He is the horse in the backyard barn and in the show ring. He is the horse running loose in the American West and is even, despite his inelegant nose, related to today's baby-faced Arabian horses. To me, he looks

* The genus *Equus*, that is.

totally practical, a horse born to weather whatever life threw at him. Like Whisper. So it seems my golden-coated Vermont horse came by his survival abilities honestly.

Injuries on the Yukon horse's body imply that a predator was involved in his death. There are teeth marks on his neck, perhaps made by a wolf. He had horsehair in his stomach. This suggested to the Yukon paleontologist Grant Zazula that he might have been wounded but had tried to heal himself by licking his injury. Maybe he stepped into a muddy wallow that, like quicksand, wouldn't let him go. I rode Whisper into a mess like that once and I had to get off just so he could pull his single-toed feet out of the bottomless mire without also having to balance me on his back.

Although it's hard for him to say exactly how the Yukon horse died, Zazula does know a lot about the world the horse enjoyed while he was alive. It was parkland. Pretty as a picture. Dotted with trees—but not forested. Something like an Alpine meadow. Maybe it even looked a bit like the Pryor Mountain peaks Jason Ransom and I visited.

It was a marvelously complex grassland. All kinds of tasty treats grew there for the horse to eat, available from the earliest spring until the dead of winter. A study of well-preserved ground nests left by a variety of critters has found thirty-thousand-year-old pollen and other remains from at least sixty different species of grasses and sedges.

For this horse, there were swaths of bunch grasses, prairie sage, wild rye, poppies, flowering chickweed . . . and buttercups. Of course, winters were tough, but the horse could still dig down through scattered embankments of windblown snow to find fresh greenery. Even today, scientists sometimes find buried bits of still-green grasses that grew at just about the time the horse died.

It was dry where the Yukon horse lived, almost arid. Precious little snow fell, and what snow did fall was light as flour. Blown by the wind, it piled into drifts. These drifts, too, helped the horse, because when the sun returned in the spring and the drifts melted, some of the meltwater sank into the soil, creating patches of green. The horse could move from patch to patch and find fresh food almost as soon as the dark time was over. These young grasses were full of protein. Jason Ransom has seen modern horses take similar advantage of long-lasting snowbanks: "The horses can move from nutrient-rich spring greens on the south faces of mountains to the same nutrient-rich spring greens on the north faces later in the summer."

Though we call this horse's home the Yukon, geographically it's considered the easternmost part of a large area called Beringia, a region that extends from the far west of the Canadian Yukon into Siberia. Over the eons, Beringia, like a slowly bobbing apple, has been sometimes partly submerged, as it is today, and sometimes above water. During the above-water periods, the region was massive, almost a thousand miles wide in some places.

In its own unique way Beringia during the ice ages was also a Garden of Eden, albeit much colder, drier, and more challenging than Messel. The Beringian lifestyle required a horse that was quite different from the little dawn horse. *Equus* was up to the task. Indeed, for the Yukon horse, Beringia may have been downright congenial—at least compared to much of the rest of the northern part of the Northern Hemisphere, where ice sheets more than a mile thick covered the land.

Horses had come a long way since the days of the Eocene, when they needed warmth and rain and grapes. The dark wouldn't have bothered the Yukon horse, because his eyes had by then evolved into huge orbs that could detect suspicious movements at great distances in very little light. His hearing was exceptional. He depended greatly on his ears, which were controlled by sixteen small muscles that allowed him to rotate them in all directions with great finesse, so that he could focus on even the slightest sounds. He could prick them forward to monitor sounds in front of him as he traveled, or he could lay them back against his head to warn companions of his displeasure. His sense of smell, already developing in the little dawn horses, was by now so well-honed that reading the aromas in the air or in a dung pile left by another horse was for him like reading a book is for us: a font of information about the world around him. His strong social instincts helped him perceive the emotions of the other band members, so that when he saw the ears of another horse pricked forward, he, too, looked up to see what danger might lie far ahead.

We won't call the Yukon horse the "perfect" horse, because evolution doesn't work that way, but we can say that his evolutionary heritage had made him a flexible, social, intelligent animal. He needed that intelligence. His pastures of plenty sat in the rain shadow of high coastal mountains, which drained all the moisture from the air. Instead, where the horse lived, there was little snow but a lot of wind. We know this because scientists have found huge drifts of loess. This light dust—like the dust that blew out of the American central plains in the 1930s and the dust that flows like rivers even today in some parts of the American West—piled up high all over Beringia.

Challenging though these windstorms were, the horse knew how to cope. This we can infer from the vast numbers of horses who lived in the region during that time. Pleistocene horse bones are very, very common. The paleontologist and naturalist Dale Guthrie ranks the horse among the "Big Three" mammals who then lived on these northernmost plains. Horses, bison, and mammoths, Guthrie says, reigned supreme in this cold environment.

So Beringia was a place to call home for many species and not an "interstate highway," as I'd been led to believe as a child. In the 1930s, scientists posited Beringia as a "land bridge" that allowed animals and people to travel from North America to Asia. The theory was all about migration. I read about the "land bridge" at about the same time I learned about the evolutionary "progress" of horses from small creatures to noble beasts. Pittsburgh, where I grew up, is a city where three major rivers come together, so I knew exactly what a bridge was: an ugly iron structure crossing water. The point of a bridge, I knew, was to get somewhere. You didn't *stay* on a bridge. Early scientists imagined Beringia the same way. They never imagined that Beringia could actually have been a focal point of evolution.

They do now, though. The geologist Robert Raynolds suggests that people think of the region not as a land bridge, but as a "food bridge," where animals found good grazing and where they could make themselves at home. As the modern world warms and the region's frozen tundra melts, so many Pleistocene carcasses have been pulled out of their burial sites that we now know that Beringia positively teemed with animals. And as those animals stayed there, generation after generation, they became ever more closely adapted to the world in which they lived.

"Evolution didn't happen in other places and things moved in," Zazula, a lifelong Yukon citizen, told me. "Evolution happened *in* Beringia. It was a really nice time to be here." So Beringia was, in its own way, a land of milk and honey.

Of course, Beringia was not Messel, with all the food a horse would need within easy reach. A horse surviving in the Yukon thirty thousand years ago would have had to be able to travel far distances for food, suggests Christine Janis. He had to have a phenomenal memory and to be able to recall the location of all water sources. He had to know when the water would be there and when the water hole would dry up. He would have had to know where the overhangs and valleys were that would shelter him from storms. He would have had to be able to flee predators when possible, and to be able to stand and fight valiantly when necessary.

And he had to be highly social. All this knowledge was too much for an individual horse to find out by himself. It had to have been passed down through the generations from the older mares to the youngest foals. If you had friends, and associated with older horses who had long memories and lots of experience, and if you were strong and hardy and cold-adapted, life in the Yukon could have been just the thing.

In the modern world, we underestimate the mind of the horse to a shocking degree. We think that because they accept our bidding and stand cooperatively, there's not much to them. But when you consider what it took for the Yukon horse to survive those winters without anyone providing grain or hay or shelter from the wind, it brings home the depth of Phyllis Preator's insight that free-roaming horses "think different."

We primates generally think of the Arctic as hostile. After all, most of our evolution occurred in the tropics. So it's not surprising that early paleontologists, who did not live in the Arctic but "heroically explored" that world (where Inuit had been getting along quite well for thousands of years, thank you), assumed that life in the Arctic was miserable. When I was a child, reading Jack London's stories of dog teams drowning in freezing rivers confirmed my worst nightmares about life up north, and I imagined animals living there as doomed to a kind of relentless Napoleonic march of misery through endless fields of snow, with a constant threat of death from starvation and exposure.

This kind of imagery drives Zazula crazy. The Yukon horse may have lived in the far north near massive ice sheets, but his world was nevertheless, Zazula emphatically believes, quite livable. Beginning about ninety-five thousand years ago, the Laurentide ice sheet began covering much of North America, spanning the continent from New England's Cape Cod in the east to the Canadian Yukon in the west. It dipped south to stop just short of where the 12-million-year-old Ashfall horses lay hidden. But most of Beringia, running about two thousand miles west of the ice sheet and into Russian Siberia, lay ice-free and above sea level.

For tens of thousands of years, the region was a kind of haven, a refuge from the relentless ice. "This special cold-arid northern grassland was like a monumental 'inner court,' surrounded on every side by moisture-blocking features: high mountains, frozen seas, and massive continental glaciers," wrote the Alaskan paleontologist Dale Guthrie.

And yet, well suited though he was to his northern world, this northern horse species, the North American *Equus lambei*, became extinct by about eight thousand years ago. Dale Guthrie wanted to know why. After examining hundreds of horse bones collected in Beringia and stored in New York City's American Museum of Natural History, he found that, over thousands of years, the northern horse diminished in size. Some scientists had suggested that human hunting had caused the extinction of the horse from North America, but Guthrie's evidence seems to point another way.

"What role did humans play in the extinction of the horses?" I asked him.

"None," he replied. "The problem with invoking humans is that you have to have some evidence." And in North America there's no evidence that humans hunted horses in large numbers.

The culprit, he proposed, was a change in climate and a subsequent ecosystem shift. When temperatures warmed at the end of the ice ages, formerly frozen landscapes became wet and in some places even swampy.

This was fine for cloven-hoofed animals like moose, but not so great for an animal that had evolved to run on a single toe. The warming climate meant that many dry-adapted grasslands disappeared, making the foods the horses relied on harder to find.

For Guthrie, his evidence shows that the northern horses were having trouble surviving in their strange new environs. "The decline in body size that I showed is a piece of evidence that the horses were having a hard time," he said. "Whatever role humans played, the horses would have become extinct anyway." Other research supports his view. The molecular biologist Beth Shapiro has looked at DNA evidence that she interprets as suggesting that the northern horses began declining in terms of population numbers as early as thirty-seven thousand years ago.

Zazula agrees. The arrival of humans in Beringia and the disappearance of the horses happened at about the same time, so that they appear, from our distant perspective, to be correlated, but that doesn't mean that one event *caused* the other. Most likely the two events are just two symptoms of a changing world, Zazula suggests, adding that the newly arrived humans may have been following the elk and moose, who also arrived in the Yukon at about the same time. All these events are symptoms of a world in flux.

"Things start to really change rapidly up here after fifteen thousand years ago," he added, referring to changes in climate that were sometimes so abrupt that they occurred within one human lifetime. "The world up here comes apart really fast. That's when things start to get ugly for the animals."

All over North America, the story was much the same.

By fifteen thousand to ten thousand years ago, North America and most of the Northern Hemisphere had become considerably warmer. We know this because, by studying ice taken from ice sheets in Greenland and in several other places, we can read the rise and fall of various isotopes of oxygen, of carbon dioxide, and of methane and other atmospheric compounds that indicate temperature fluctuations. The more methane in the atmosphere at a particular time, for example, the hotter it was.

It turns out that what most of us know as "*the* Ice Age," once imagined to be a world eternally encased in a vast field of ice, was not an Ice Age at all—but a *series* of ice ages that caused the ice sheets to ebb and flow over the Northern Hemisphere with surprising alacrity. Near the end of the Pleistocene epoch, temperatures sometimes rose and fell so abruptly that one team of scientists has likened the climate to a "flickering switch."

The northern ice cap began melting in earnest about 14,500 years ago, but even this wasn't one long, smooth process. The ice disappeared slowly in some regions, but more quickly elsewhere. There was no uniformity. Instead, the melting was erratic, occurring in fits and starts. Glaciers would melt back, then another cold spell would occur. Because of this instability, plants that grew in an area might not be able to grow there in following years. Ergo, the animals depending on the plants would not be able to find them where and when they expected them.

The melting also brought a whole new array of insects to harry the animals. Indeed, scientists study temperature fluctuations in part by finding fossils of various species of beetles, which are closely adapted to temperature. This means that beetles, of all things, can provide clues to ancient climate, just as ice cores can. By knowing which beetle species live at which temperatures, the temperature of a particular area can be estimated for a given time frame.

All this evidence points toward a significant climate event that may have contributed to the demise of the North American horses, both in the Yukon itself and throughout the whole continent. About 12,900 years ago a global chill, called the Younger Dryas, dramatically reversed the warming trend. Then about 11,500 years ago the temperature suddenly jumped back up by as much as 8 degrees or so Fahrenheit in some places over the course of only about sixty years. The Younger Dryas was like the Eiffel Tower of Heat that began the Eocene—but in reverse. First it got cold, then it got warm. This warming was not globally uniform. Pollen studies show that the Amazon experienced only minor blips and that, even in the north, temperatures remained stable in some refuges.

Various species were affected differently by this anomaly. At least one species—*Homo sapiens*—adjusted by making a major lifestyle change:

Archaeologists widely attribute the beginning of farming to the warming and drying of parts of the Middle East, where people had long relied on hunting and gathering. When the number of wild animals began to diminish because the temperatures changed, we had to learn how to domesticate them. We also learned how to raise crops and, to accomplish that goal, began learning how to control the flow of water by digging rudimentary canals. Clearly, the change in temperature was putting a lot of pressure on human beings. The same changes would also have pressured other living things.

The cause of the warming is under debate. One theory with considerable traction suggests that so much freshwater was released from the North Atlantic ice sheet into the northern Atlantic itself that global ocean currents shifted, changing rain and wind patterns. Another theory proposes that the melting ice sheets released large numbers of icebergs that drifted as far south as the Iberian peninsula, disrupting ocean currents. In support of this theory, marine scientists mapping the deep ocean floor have found what seem to be mounds of otherwise unexplained stone rubble. The theory suggests that this rubble dropped onto the ocean floor out of the melting icebergs.

Unfortunately, when considering the Pleistocene extinctions, our very refined data is limited to Greenland and Antarctica and to parts of Europe. We don't know precisely what happened to temperatures in most parts of North America. We just don't have enough detail.

We do know that instability was rampant. In *The Great Basin*, the archaeologist Donald Grayson documented ongoing chaos in the interior of the American West during the millennia between fifteen thousand and ten thousand years ago. The extent of the chaos was biblical. Lakes rose to high levels, causing massive floods. Some, like the Great Salt Lake, rose and fell, and then rose and fell again. At times these regions would have been flower-filled, at other times they would have been inundated. At still other times, they would have been deserts.

While available evidence now shows that *Equus lambei* persisted in some far north locations until at least eight thousand years ago, horses in the rest of North America had become extinct by about eleven thousand years ago. In some specific localities, extinction does appear to have occurred abruptly. One such place is the La Brea tar pits in Los Angeles. At this site, a *Lagerstätte* of southwestern North American coastal life, horses were plentiful about forty thousand years ago. Then, about eleven thousand years ago—they disappear. "First they're here—and then they're not," the paleontologist Eric Scott told me when I visited La Brea. "And after the horses vanish, we don't see people in this area for another two thousand years."

The La Brea research site, open to the public, is unique. It's located right in the center of urban Los Angeles, near the city's expensive clothing stores, high-end art galleries, and museums. Just a few hundred feet below these buildings is the Salt Lake Oil Field, a subterranean pond of thick, sulfurous oil that slowly seeps up to the surface from the depths of the earth. The upwardly oozing oil from this field creates puddles of thick tar that, on a hot day, are sticky enough to trap someone, or something, foolish enough to walk over it. It's like nature's version of flypaper. Even today these puddles, camouflaged by a covering of leaves and dirt, can cause problems. If you're not careful, you might easily step on what seems like solid earth, only to find your feet immersed in oil. Every once in a while, La Brea staff must rescue some hapless creature, like a squirrel, who has become hopelessly mired.

According to scientists, that's what happened to a lot of animals beginning about forty thousand years ago. Unlucky horses, bison, and many other species, caught in these traps, were not able to escape, and died there. Some animals, including a lot of dire wolves who preyed on the trapped animals, sank into the seeps and were forever entombed—until paleontologists found them, pulled them out, cleaned them up, and studied them.

When scientists began researching here a century ago, they found so many bones that they established a permanent research laboratory at the site. At first, paleontologists were interested in the larger animals, but now they also study tiny seeds and other bits of plant life, so that today they have a good understanding of the complete ecosystem in which the horses lived.

This is laborious work. I watched one volunteer patiently using a microscope and forceps, examining some of the tar, separating various plant seeds and detritus into individual piles for further study.

"How much time do you spend each day doing this?" I asked.

"About eight hours," he answered.

"Do you get tired of it?"

"Never," he answered.

I thought of Chris Beard searching for ten years in poison ivy and kudzu to find one horse tooth, and of Matthew Mihlbachler measuring seven thousand fossilized horse teeth, and realized once again that paleontology is only for the stouthearted.

Evidence from La Brea researchers implies that, at least locally, the disappearance of horses is not connected to the presence of humans. Researchers have discovered only a single human skeleton there. That skeleton dates to about nine thousand years ago, several thousand years after the date of the most recent horse skeleton. Scott, who is an expert in equine paleontology, cannot explain why horses disappear from the local record about eleven thousand years ago. There's no evidence from the horse skeletons that the horses were suffering in any way, no evidence that they were getting smaller, as in Beringia, or that their bones were brittle because of some new disease. They just vanish.

This extinction event was not limited to horses. Other large animals were also disappearing. Across North America, an estimated 72 percent of the large-bodied mammals died out. During what's formally called the Quaternary Extinction Event, North American tapirs disappeared, as did short-faced bears and Florida cave bears, giant ground sloths, giant beavers, camels, saber-toothed cats, dire wolves, mammoths, and mastodons, along with plenty of other large mammals.

Scientists have been unable to pin down the cause of the extinction of horses and other animals. In fact, the issue is surprisingly contentious, even vituperative at times. In the 1960s, the geoscientist Paul Martin asserted that "man and man alone" was responsible for the horses' disappearance, by overhunting them. Others claimed that climate change was the culprit. While trying to sort out the various points of view, I found myself becoming interested not only in the mystery itself, but in the mystery of why the battle was so bitter.

To better understand, I visited Larry Agenbroad, then director of the Mammoth Site in Hot Springs, South Dakota, where large numbers of young male mammoths fell into a sinkhole twenty-six thousand years ago, died, and were preserved. Other animals, including fish, frogs, birds, rabbits, squirrels, wolves, coyotes, and even a llama, have also been found in this hole.*

Horses are conspicuously absent. Agenbroad told me that horses were too smart to have been trapped this way. I liked the explanation, but had to be honest: Given the young male horse behavior I had seen in Wyoming, I didn't entirely buy his theory. Young stallion behavior can be pretty risky. Indeed, while Jason Ransom was studying horses in Colorado's Little Book Cliffs, a band stallion turned up dead at the bottom of a cliff. Tracks at the top of the cliff showed there had been a violent fight. "One of them didn't judge too well," Ransom commented.

Agenbroad is on record as firmly believing that humans helped exterminate mammoths from North America, but he disagreed with the theory that humans exterminated the horses.

I asked him why.

"There is no evidence anywhere of humans hunting horses," Agenbroad replied, referring to North America. He was almost shouting. "We don't have a good bulletproof theory for their disappearance," he said. "I think that horses were pretty well gone by the time the people got here."

But why? I was insistent.

He answered with a shrug of his shoulders, and then asked his own question. "We've gone through worse and better climates and they survived. Why did they die just when a big smorgasbord was being laid out for them?"

So it was back to square one.

When Paul Martin first postulated in 1967 that early humans had engaged in a veritable killing frenzy, he chose the unfortunate, loaded word "blitzkrieg." Today that word wouldn't raise many eyebrows, but

* Sadly, Agenbroad died shortly after our conversation, at the age of eighty-one.

in the years following World War II, "blitzkrieg" reminded people of the relentless bombing of London and other cities. His word choice implied that the Clovis people, who arrived in the New World with revolutionary hunting technology around twelve thousand or so years ago, coincident with the Younger Dryas, engaged in senseless, bloodthirsty violence. Their advanced killing tools, Martin proposed, made mass slaughter of horses and other animals a likely outcome. This did not sit well with many people, including native North Americans. Discussions quickly became bitter.

Essentially, Martin was starting a rumor. Although he had a clear correlation—humans arrive; large animals disappear—he had no real evidence of direct cause and effect. As I searched the literature, I found that Larry Agenbroad was essentially correct: there are no North American sites showing that humans engaged in large-scale horse slaughter. There are, however, a few sites potentially indicating that people did hunt horses when they found them, but these sites are not widely accepted.

For example, at Oregon's Paisley Caves, the archaeologist Dennis Jenkins has found what he thinks is evidence—human coprolites (fossilized feces)—that humans may have been there very early, about 14,300 years ago. He's also found a few horse bones in the area, which would have been marshy wetland at the time, along with what he says are human-made stone tools with horse-protein residue. But Jenkins's findings are questioned by other scientists who say that his dates may be incorrect.

When I talked with Jenkins about his research, he told me that although the region is dry now, in those days it would have been great horse country: "There would have been stands of grass. In other places, slightly more elevated, there would have been junipers and ponderosa pines in patches. About a mile away from the cave was the edge of a lake, fairly deep, that likely would not have been frozen over often. There was a marsh where the river dumped in. You'd see all that from the caves. You would certainly have seen mammoth out there. Horses. Camels. The occasional carnivore would have included the American lion."

Jenkins suggests that these early people, whoever they were, did not stay long. He doesn't know whether they were hunting the horses or scavenging them. Either way, if you accept his evidence (and again, not all researchers do), it would appear that humans and horses did coexist here for at least a short period of time, several thousands of years before the Clovis people began showing up. Even if that turns out to be the case, though, the small number of horse bones present points toward the theory that horse populations were dwindling long before people became numerous enough to cause the horses' extinction, since only a few bones have been found.

At another site, Wally's Beach in the Canadian province of Alberta just north of the Montana border, Brian Kooyman and Len Hills have studied thirteen-thousand-year-old bones from seven horses which, they say, were butchered by humans. These seven skeletons each lay at a distance of several meters from the others, which Kooyman and Hills interpret as indicating seven separate killing events, each of only one horse. The researchers also found what they believe to be tools. Analysis seems to show that horse proteins were on the tools, more clearly than at Paisley Caves. Kooyman and Hills go a step further than the Paisley Caves scientists by suggesting that humans were not just hunting horses, but *preferentially* choosing horse meat over the meat of other animals. This site has preserved tracks of other animals, but no evidence that any other animals were butchered. Their theory is interesting, but again, other archaeologists are skeptical of the team's conclusions, citing the difficulty of accurately dating the evidence.

There are several other North American sites like these where horse bones appear to exist in the same layer of time as human artifacts, but all are disputed. One thing *is* clear, though: no horse-hunting sites in North America have been found that even begin to approach the scale of horse-hunting sites that exist throughout Europe and Asia. If horses *were* hunted by early North American people, the scale of the hunting doesn't seem to merit Martin's "man and man alone" assertion of "blitzkreig."

Nevertheless, his theory has persisted. I'd heard this idea so often that I'd always assumed it was correct, so when I discovered that it was little more than a myth, I wondered how it got started. I decided that when Martin first proposed his idea, Western civilization was suffering from two crises of confidence—the worldwide economic depression of the 1930s, followed by World War II and the terrifying power of the atomic bomb. The Victorian story of humanity's onward-and-upward progress had ended with World War I. Then, following World War II, we entered a period of collective self-loathing, so that when Martin suggested that a group of a hundred people arriving in the New World about twelve thousand years ago not only could—but *would*—lay waste to an entire species, people listened.

Today, views like Martin's have been moderated by scholars. The word "blitzkrieg" has been dropped. In its place, the commonly heard term is "tipping point." In this modernized view, which takes into account the accumulating evidence of disastrous climate change following the melting of the ice (information that was not available to Paul Martin when he proposed his theory), humans were not mindlessly committing universal mass slaughter. But, some scientists suggest, humans *did* play a role as a newly arrived predator.

Humans were "the *necessary* tipping point," the archaeologist Stuart Fiedel told me: without humans added to the mix, the animals would have survived. Fiedel's colleague Gary Haynes believes that long before humans arrived on the scene, the Pleistocene horses had been contracting into ever-smaller ranges, so that small-scale human hunting might have been just enough to push the horses over the edge. Haynes has even taken a stab at guessing how many horses were living in North America when people arrived: only 1.2 million. That's not very many, given the size of the continent.

So the question remains: Why would horse numbers have been so severely diminished?

When thinking about cause and effect, philosophers often use the terms "proximal" and "distal." The proximal cause of a forest fire, for example, might be human behavior: a campfire left burning. But the distal cause of that same forest fire might be a longer-term factor or factors, such as drought. Likewise, the proximal cause of the catastrophic floods in Colorado in 2013, which rose all the way up to the edge of Jason Ransom's family homestead, was obviously an unusual deluge lasting for days. A distal cause might be that ongoing drought had killed vegetation that might have encouraged some of that moisture to soak into the ground rather than run down the mountainside. And an even more distal cause might be that our shifting climate has caused changes in global wind patterns, which brought Gulf of Mexico moisture all the way up to High Country desert.

Your interpretation depends on how far back you want to follow the chain of cause and effect. Given our improved understanding of late Pleistocene temperature and climate, the list of proximal and distal causes for the great extinction mystery is growing on an almost daily basis. For example, the plant paleontologist Jacquelyn Gill has found that, after a system of plants becomes fundamentally degraded, it will likely take hundreds of years for a different ecosystem to become firmly established. In the meantime, there may well be a comparative paucity of vegetation varieties.

The smaller mammals might be able to find their way through the morass, but, suggests the evolutionary biologist John Tyler Bonner: "Size matters." Bonner explained the extinction of large mammals to me this way: "Most evolutionary biologists would say that this [occurs more easily] because there are fewer of them. There are many more rodents than there are elephants. The chances of becoming extinct because of a natural event is more likely with large mammals. All you have to do is knock out a few elephants. Remember: size limits the abundance of species. If you have fewer around, then it's easier to make them extinct."

I suggested that this might be particularly true of an animal like *Equus*, which only reproduces once a year. A large animal with limited reproductive abilities would certainly be vulnerable to quick shifts in vegetation and climate. And a large *nonmigrating* animal would have been even more vulnerable. We don't know for sure whether Pleistocene horses migrated or not, but we do know that most modern horses stay within a home range. Perhaps this affinity for a home range also made them more vulnerable to ecosystem changes. And perhaps as the world changed, each of the isolated populations of horses became smaller and smaller, until *Equus* simply vanished from the continent where he and his ancestors had lived for 56 million years. Like dust in the wind.

INTERMEZZO

One spring Vermont day when I still had Whisper, the weather was glorious. Windmill Hill, the dirt road leading to my house and barn, was knee-deep in mud. A few piles of snow still dotted my yard, but the sky was an inviting soft blue. A tempting smell of warmth wafted through the air. Tender young leaves were budding on the sugar maples, and patches of grass and curlicues of young ferns pushed up along the road's shoulders. It was the kind of day that made any heart glad.

Whisper's gladdened heart got the urge for going. I never saw him leave, or Gray follow him. But yielding to some gnawing sixth sense, I got up from my piano and looked out at the paddock for the horses. Nothing. They could have been standing in their barn, since I'd left the barn door open for them. But I doubted that. It was just too nice a day, the kind of day that Whisper was likely to be very pleased about.

I walked down the hill to check. Nothing. I walked over to my neighbor's front yard. Nothing. I checked all the usual places, walking about a mile down the road. No horses. But then I saw their tracks. It looked as if the two were headed all the way down the hill to the school yard. Their tracks were quite directed. They were not meandering. They had a definite goal in mind.

I panicked. Who knew what these guys were up to? In those days, I didn't know that horses rarely go far. I imagined them ending up in New Hampshire, or even Maine. At a fork in the dirt road, the tracks changed direction, turning right. Now they were heading away from the school and down an overgrown cart path, abandoned for more than a century, since the Green Mountain hill farmers headed west looking for better wheatlands. Vermont has an awful lot of forest laced with these old roads. If you know them, you can travel all the way from the state's border with Massachusetts up to Canada. Great for riding. Not so great if you're chasing horses on the move.

To find a horse, ask a horse. I borrowed one of my neighbor's horses, threw a bridle on her, and let her decide. She moved with purpose. She, too, knew just where she wanted to go. She chose yet another old cart road, one that quickly became little more than a footpath, but I knew we weren't just wandering in the wilderness because the path ran along one of the state's ubiquitous tumbledown rock walls, evidence of long-abandoned farms. I wasn't on a road to nowhere. I was going *somewhere.* I just didn't know where. It was a nice day for a ride, but as we moved ever more deep into the forest, my anxiety increased.

Then, suddenly, there they were. Muzzles down, lips pulling away at the nicest patch of grass a horse could imagine. Perfect grass. Grass that must have been as sweet to them as Dutch chocolate is to us.

The golden palomino and the gray Percheron stood in rays of bright spring sunshine. My Tom-and-Huck duo looked angelic, almost as though they were surrounded by a halo in the midst of the dark evergreen forest. Beatific. Innocent. Not at all like criminals.

Above them was that soft blue sky.

I was awestruck. Evolution had blessed them with some powerful survival tools. How in the world had they found this particular patch? What was built into their senses and their brains that led them out into the middle of a forest to the one place where grass grew?

In those days I had no idea of their deep history, of the events that brought these horses into my care. I knew nothing of the Cretaceous Terrestrial Revolution and the 100-million-year-old emergence of flower power; of the evolution of grasses some 50 million years later or of their gradual spread worldwide. No one had told me anything about the constant and sometimes violent shifting of the tectonic plates that float atop our boiling-soup-kettle planetary interior; or of the sudden chill 34 million years ago; of the extreme heat of 56 million years ago; or the ups and downs of carbon dioxide; or of the near extinction of all horses during the days of *Epihippus*; or of the Yukon horse; or of Charles Darwin; or even that horses had become extinct on my continent long ago.

But young as I was, I knew even then that Whisper had led me to one of my life's best memories.

6

THE ARCH OF THE NECK

The steady thunder of horse hooves echoes deep within the human soul.

—TAMSIN PICKERAL, *The Horse: 30,000 Years of the Horse in Art*

Sometime around 66 million years ago, Spain began attacking France, tectonically speaking. The collision had been a long time coming. Since the days of the Cretaceous Terrestrial Revolution, the Iberian tectonic plate had been slowly creeping northeast, toward the Eurasian plate. The two finally clashed at the end of the dinosaur era. Since then, in a David-and-Goliath shoving match, the tiny Iberian plate has pushed relentlessly against Europe, which has stood its ground, backed by the immensity of Asia.

The offspring of this battle is the Pyrenees, a roughly three-hundred-mile-long rugged mountain range topped by windswept and nearly impenetrable peaks and filled with deep fertile valleys. The Pyrenees run from the Mediterranean to the Bay of Biscay on the Atlantic coast. On a topographic map, the entire range looks like a poorly stitched seam patching Iberia onto Europe.

In reality, though, since the days of the ice ages, these mountains have stood as a sheltering bulwark, protecting living things—including horses and humans—from many of the ravages of cold, snow, and ice. This is why Iberia is said to have been a refugium—a haven—from the ice sheets of northern Europe.

The rising of the Pyrenees was a stroke of good luck for the advancement of human civilization and for the partnership between horses and humans. The mountains created a sheltering infrastructure in which large numbers of people could live comfortably. Geological forces associated with the mountains' rise carved out interior caverns and rock shelters that were ready-made living spaces, where families could stay for months at a time in close proximity to each other, sharing food, work, and ideas around community fires. Archaeologists have found evidence of such community hearths, around which artisans left such huge piles of debris from their carving of stone and ivory that researchers today, tens of millennia later, can still see exactly where each person sat.

Historians use the term "culture hearth" to describe a place—Mesopotamia, for example—where, for whatever reason, the stars are so well aligned that an immense leap forward in human thought can occur. Culture hearths provide what we today call "connectivity"—the ability to interact with people outside the family. And long ago, the naturally created infrastructure of the Pyrenees region hosted an extensive connectivity—real hearths that nurtured creativity and intellectual advances.

In the form of artifacts we have plenty of solid evidence of this extraordinary culture, showing us that the people's basic needs were so easily met that they were free to devote much of their energy to the expression of beauty. There was even music: flutes made from the hollow bones of birds have been found in some of these shelters. It was yet another halcyon time, like Messel.

This time, though, the keystone to the nurturing natural system was not thick foliage and ripe fruits, as at Messel, but grass. Grasslands were prolific, while forests were limited. Water rushing off the mountains carried rich topsoil into the river valleys, nourishing grasses and other plants that fed reindeer, horses, mammoths, and other animals. These river valleys were the people's pantry—a general store, if you will, a ready source of food where the shelves were never bare.

Before I visited the region, I imagined Pleistocene people as desperate hunters, wandering aimlessly through a frozen world looking for food. The truth is the opposite. Despite the icy temperatures, these

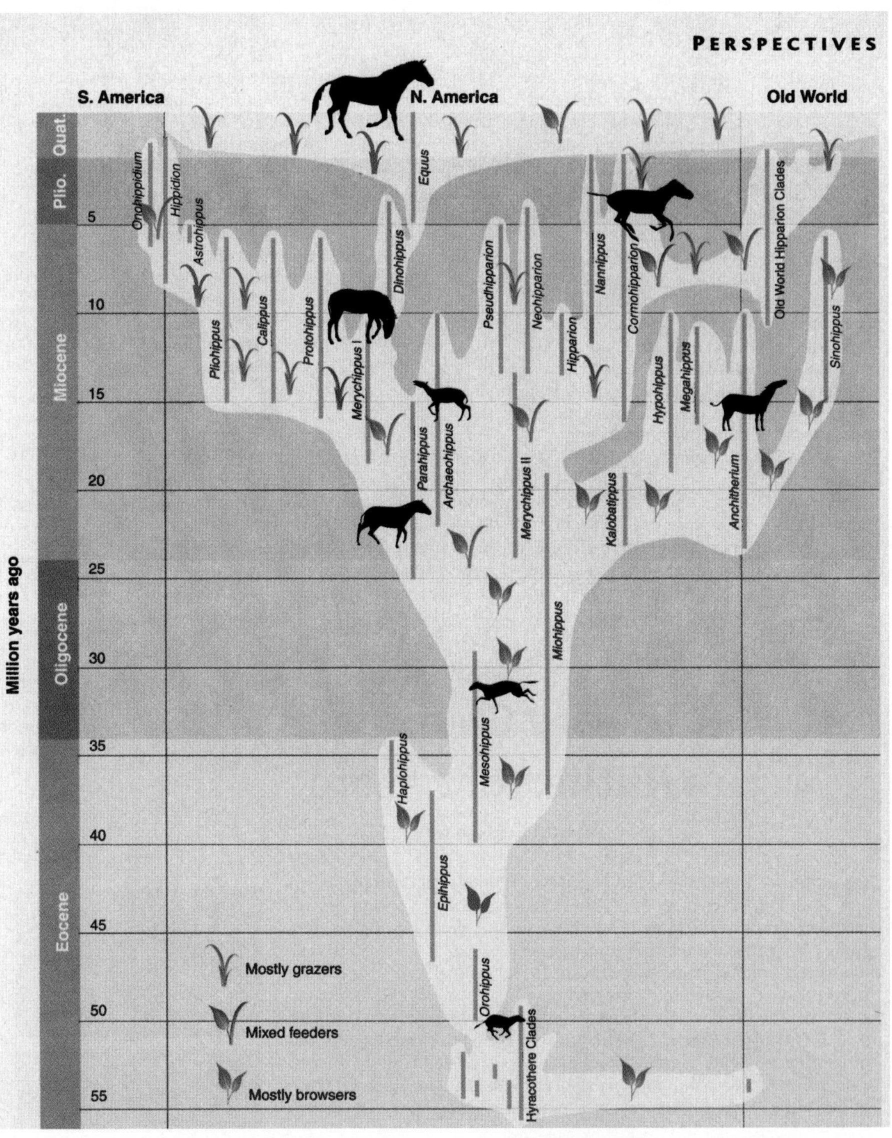

An evolutionary tree of the horse's genetic family, Equidae, showing changes in geographic distribution, diet, and body sizes over the past 56 million years

(From Bruce J. MacFadden. "Fossil Horses—Evidence for Evolution." *Science* 307 (2005): 1728–30. Reprinted with permission from AAAS)

TOP: ***Three mustang stallions at McCulloch Peaks, not far from Cody, Wyoming*** (Greg Auger)

ABOVE: ***Watching Pryor Mountain mustangs with the ethologist Jason Ransom*** (Greg Auger)

Horses can thrive in both the wettest of climates, like the hyper-Atlantic coast of Galicia, Spain, and extremely dry regions, like the Bighorn Canyon National Recreation Area in Wyoming, U.S.A. (Greg Auger)

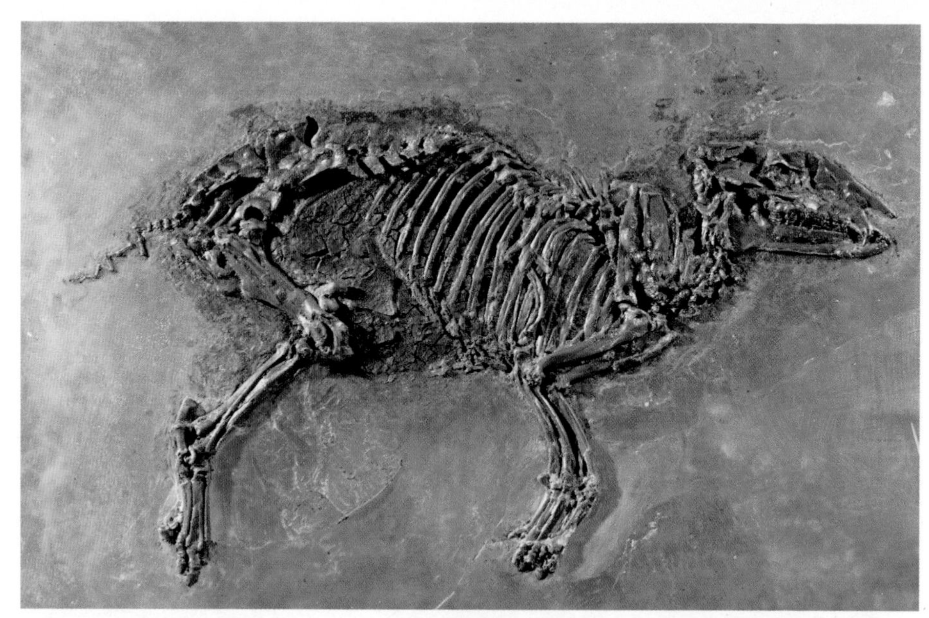

A 47-million-year-old dawn horse preserved at Messel (© Senckenberg, photograph by E. Haupt)

Stylized horses painted more than twelve thousand years ago on the walls of the Ekain cave, a UNESCO World Heritage site located in the Basque region of Spain (Jesús Altuna)

Herwig Radnetter and his Lipizzan stallions (Greg Auger)

The intelligence and sensitivity of horses has been an ongoing artistic theme for at least 35,000 years, seen here in Sawrey Gilpin's Gulliver Addressing the Houyhnhnms, *1769.* (Yale Center for British Art, Paul Mellon Collection)

Diego Velázquez's Philip III on Horseback, *1634–1635* (Prado Museum), *and a several-thousand-year-old petroglyph of a horse and rider from the Campo Lameiro in Galicia* (Greg Auger). *Both Philip III and the ancient rider are depicted holding what the equine ethologist Laura Lagos calls the "stick of power." This kind of imagery—of horse, rider, and weaponry—has served as a symbol of power for thousands of years.*

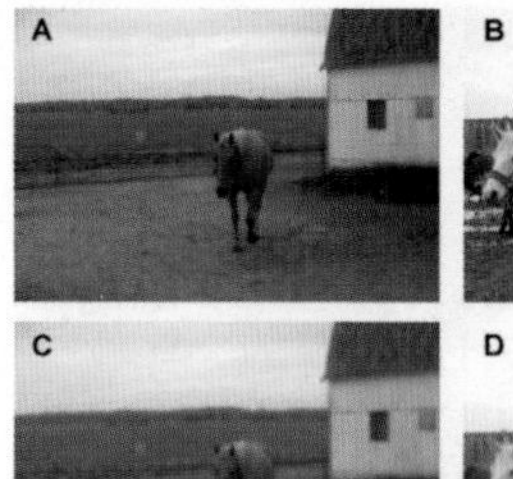

The colors seen by horses are much more limited than those seen by humans.

Human Trichromatic Color Vision

Horse Dichromatic Color Vision

The difference between the color vision of a horse and typical human color vision (From Joseph Carroll. "Photopigment Basis for Dichromatic Color Vision in the Horse." *Journal of Vision* 1 (2001): 80–87)

ABOVE LEFT: ***Lukas, owned by the Californian Karen Murdock, poses with his Guinness World Records certificate, awarded to him for "the most numbers identified by a horse in one minute."*** (Joan Malloch, courtesy of Karen Murdock)

ABOVE RIGHT: ***Horses have a horizontal visual streak that allows them to see almost directly behind them. Blinders help keep horses in harness from becoming frightened by the vehicles they're pulling.*** (davidelliotphotos / Shutterstock)

Kris Kokal and his family rehabilitate mustangs at their farm in New Hampshire. Here, Kris and Belle, one of his rescued horses, enjoy a snowfall. (Greg Auger)

LEFT: ***A Mongolian herder with his horse*** (Greg Auger)

BELOW: ***Inge Bouman speaks about rewilding the Takhi at the press conference in Ulaanbaatar celebrating the twenty-year anniversary of the return of the horse to Mongolia.*** (Greg Auger)

BOTTOM: ***Young Mongolian jockeys await the start of a traditional horse race.*** (Greg Auger)

Pleistocene people lived in a comfortable world. In what would become France, people lived in close proximity under rock overhangs above wide river valleys. In Spain, they lived high above grazing lands in the mouths of massive caves.

These overhangs and caves, carved out by water rushing down through the limestone from the rising Pyrenees, created "neighborhoods." You didn't have to live alone. Your friends and family could set up house only a few minutes' walk away in yet another overhang or cave mouth. This was luxury living, sometimes complete with penthouse suites and balconies.

These sites were permanent, but people did not live permanently at any one site. As hunter-gatherers they moved in a yearly rhythm, from one place to the next—depending on the patterns of nature. But their lives were not haphazard. Just as we return year after year to our favorite summer spots and winter retreats, they returned to the same familiar places season after season.

Pleistocene neighborliness coupled with a rich natural system allowed for the kind of stability that nourishes human culture. Comfortably sheltered and well fed, people had leisure time. They created art, and even fashion, out of whatever was available in the world around them. They carved beads by the thousands to adorn their clothing, using the sinews of horses to sew together those finely tailored costumes. The tusks of mammoths were carved into eyed needles. In one museum I visited, signage explained that these sewing needles were better made than those made by artisans of the Roman Empire. These Pleistocene artisans shaped flint knives as sharp as modern steel knives. I'd always imagined that stone tools were "primitive," but in France I watched a modern flint knapper make these blades using traditional technology. Touching the knife he had carved, I nearly cut my finger.

In the region north of the Pyrenees, the residences of these people were sometimes quite sophisticated. They often chose living sites with southerly exposures where the sun could shine in. They devoted a lot of time to modifying these spaces. Under some rock overhangs, researchers have found rings carved into the natural stone ceilings that, some scientists speculate, may have been used to hang hides that made walls to protect people in foul weather. Sometimes, archaeologists suggest, these hanging hide "curtains" may even have been used to divide a space into separate rooms. Archaeologists point to clues like these when they suggest that the same people returned year after year. People just passing through would not take the time to create such elaborate arrangements.

On the other hand, the Iberian residences reminded me of modern high-rise condos: one mountain I visited had many caves and cave mouths, and archaeologists have found evidence that many of these mouths were occupied.

In France, the residences, called *abris*, extend for miles along river valleys, like suburban developments. The neighborhood kids probably played together. When the adults got together, maybe they indulged in the Pleistocene equivalent of quilting parties or knitting evenings, or maybe they told stories for hours around the fire about the one that got away.

Of one thing I am certain: many of those stories had to have been about what the horses were up to that day.

One thoroughly wet, chilly day in late May 2013, I walked up a gentle, narrow stream valley that fed into southwestern France's Vézère river. This small vale, called Les Roches, near the village of Sergeac, is like a suburban cul-de-sac. The Vézère, a major regional waterway, was a superhighway along which people could easily travel. Many Ice Age people lived along the Vézère's high cliffs.

But the small side valley I was walking up was in these days much more intimate. Researchers have found a number of home sites here. Some were occupied thirty thousand years ago, while others were used only a few hundred years ago. Given its long-term popularity, Les Roches must have been prime real estate. As I walked around, I could see why. The Vézère valley is wide and easily accessed by invaders, but this little side stream offered privacy and protection. The streamside habitat created a ready-made hamlet that seemed rather cozy. I could easily imagine comfortably camping here for quite a while.

I was meandering up the vale on a whim, as I'd heard a scientific lecture in New York City months earlier about archaeological artifacts that had been found there. I had no particular goal in mind—I just wanted to see the place. What was it like to live on a Pleistocene side street?

But I lucked out. I encountered Isabelle Castenet, the descendant of a man who helped discover the ancient living sites here. Her family had begun to excavate in the late 1800s, and the family still has strong regional roots.

Castenet, an expert on local history, was happy to chat. She maintains a small museum, full of artifacts discovered by her family members over the past century. Not only was I fortunate enough to encounter her, but, since it was a lousy day and hardly anyone was around, she had plenty of time. Hoping to get a sense of whether the fascination that people elsewhere had for horses held true even here, in this unprepossessing place, I asked her to describe how the people of Les Roches lived.

"During the Pleistocene, there may have been as many as a thousand or even two thousand people living in this whole area at times," she said, waving her hands out toward the Vézère. In her view, the people probably rarely fought, since they had no real property of value, other than what they made with their own hands. Their survival, she added, probably depended on mutual cooperation. She imagined them living together in harmony and trading work and art from one living site to the next. Her exotic descriptions reminded me of some of the back-to-nature landscapes of the late nineteenth-century French artist Henri Rousseau.

She explained that people here ate mostly reindeer, the region's most common animal at that time. Curiously, Castenet said, despite this reindeer diet, many of the caves in the region contain a lot of horse art. She showed me some of this art, as well as some images of horses that were etched into bones and onto stones.

Charmingly, the horse art isn't very good. The artists in this small vale created lots of decorative images, but much of the work is rough, almost childlike. They reminded me of the art our own modern children like to create. The various body parts of the horses don't fit together.

The proportions aren't right. One horse has a head so heavily indented that his nose looks a bit like a crocodile snout. I won't say that these artists drew animals like I draw animals (few people are that inept), but the images have none of the astonishing polish of the Vogelherd ivory carving or of the Chauvet Cave paintings.

I was enlightened and fascinated: Then, as today, artistic genius may have been relatively rare. Those with the talent of a da Vinci or a Rembrandt created Pleistocene masterpieces that still thrill us in modern times. Others were less talented but still valued the experience of creativity enough to engage in it and to leave behind a few crude lines etched into stone.

There are many such sites in France, along with the other, more famous sites that contain the most spectacular art. Not far from the Vézère, there's an electric-powered, open-topped little train that carries scores of tourists about a half mile down into the depths of a cave called Rouffignac, containing numerous masterpieces with many horses. In the replica of the Cave of Lascaux (the real cave, damaged by decades of tourism, is now off-limits to the public) there are joyful bands of Pleistocene horses on the cave walls.

But of all the caves with horses I visited in France, I most treasured those shown to me by Isabelle Castenet on that chilly afternoon, perhaps because while talking with her I finally understood: the people who lived in this little side valley, on their own home street, were quite like us. They engaged in the mundane activities of quotidian life, ate and drank, carved tools, sewed clothes, and in their spare time, they drew what they saw around them. Even if they weren't particularly talented, they left behind a record of what they saw and of what was important to them. And from this artistic record, it's apparent that what they loved most was the beauty of the natural world in which they lived.

And the record that they left shows how much they revered horses.

Spanish Pleistocene art also reveres the horse. Where the French living sites were strung out along rivers, the Iberian sites are often clustered on top of each other. For example, Monte Castillo is a tall, cone-shaped mountain that stands high above the small Cantabrian village of Puente Viesgo, located just a bit west of Bilbao on the northern Iberian coast. Surrounded by three rivers and grass-filled valleys, this stand-alone towering peak reminds me of a very large Egyptian pyramid.

This mountain contains many caves and cave mouths. Archaeologists have found evidence that many of these cave mouths were occupied by people, so that again, this must have been coveted real estate. With commanding views of the grasslands in the river valleys below, you could have kept close watch on the wildlife and ordered up for dinner. In France, most of the caves are long and very narrow, but along the Spanish northern coast underground rivers have carved huge interior spaces that could hold as many as one hundred or more people. In France, people could not gather in the caves, but in Spain these caves may well have served as "community halls"—gathering spots where information could be shared and ideas exchanged. Where the French caves were tight, oxygen-starved spaces, Spanish caverns were immense and downright welcoming.

Welcoming too were the Spanish guides who led me, along with other visitors, through those caves that were open to the public. Touring the popular French sites can be a hassle, particularly in the summer when lines just to buy tickets may require hours of waiting. In comparison, Spain was peaceful. Fewer people visit the Spanish sites (the French government works hard to market to tourists), but the art is equally compelling.

In one large Monte Castillo cave, the Spanish guide told me that Pleistocene people came into the mountain to dance together. (Maybe. Since I didn't speak Spanish, I couldn't question her more closely, but I noted her words with an asterisk that reminded me to check up later to see if this was local myth or scientifically documented fact. I found no scientific paper to confirm her statement, but I thought of Edvard Grieg's "Hall of the Mountain King" and enjoyed the image.)

The archaeologist Lawrence Straus, who has devoted his career to studying Iberian Pleistocene sites, says the larger caverns do show evidence of heavy human use. Straus and his colleagues have found in some of these caves an abundance of butchered horse bones. The people of the French Dordogne seem to have eaten horse meat only occasionally, but Straus and friends say that in Spain Neanderthals and the first *Homo sapiens* relied heavily on this source of food. The difference may have been due to the greater number of horses then living in Spain as compared to France, but it may also have been due to the lack of reindeer herds in Spain.

Straus and his colleagues have also found that the region's Neanderthals and *Homo sapiens* may have hunted horses using different techniques. By analyzing bones left in various Monte Castillo caves, these researchers have found that the Neanderthal inhabitants hunted individual horses and brought parts of the carcasses back to their living sites for further processing. *Homo sapiens*, on the other hand, hunted larger numbers of horses and butchered the meat in the field, hauling back only the choicest pieces. These differences might be due to the differing hunting abilities of Neanderthals and *Homo sapiens*, or it could be because there were more horses available to hunt, or it could just be due to a difference in food preference.

Nevertheless, these findings do tell us something important: although no evidence of large-scale horse hunting exists in North America, horses in Iberia and elsewhere in Europe and Asia *were* commonly hunted, sometimes in very large numbers. Yet on these continents, horses did not become extinct.

There were plenty of horses in the region thirty thousand years ago. The northern Iberian coastal plain, the littoral, was then much more extensive than it is today, because so much of the world's water was frozen. Remarkably, this littoral was biologically the westernmost part of the Eurasian steppe, which stretched all the way from the Atlantic Ocean to Siberia. The Iberian horses then living near Monte Castillo would have enjoyed complex grasslands that were somewhat similar to the grasslands enjoyed by Grant Zazula's golden-coated Yukon horse. Perhaps they even ate an Iberian version of buttercups.

Equus thrived under these circumstances. The Messel horses were island animals, but during the Pleistocene, Eurasian horses had a grass-filled world at their command. It was as though this world had been made for them. It was during this time that the Vogelherd horse was carved, and during this time that one of the most astonishing and mysterious masterpieces of Pleistocene cave art was created.

This masterpiece, Chauvet Cave, shows horses playing a unique role as nervous observers of the wildlife around them. Chauvet sits high above a river valley, and thirty-thousand-plus years ago, this valley was rich with life. Few researchers or scientists are allowed to enter Chauvet, but there are many books and videos available, so that it's possible to experience the cave's art virtually.

The paintings show lions, reindeer, bison, bears, aurochs (a form of early cattle), a large panther-like cat, a megaceros (a large deer), ibexes, owls. The scope is immense, so much so that researchers at first doubted the authenticity of the cave. It didn't seem possible that "primitive" people could have created such a panorama.

In this raucous panoply, the horses are essential. They are watchers. Many of the Chauvet animals are shown interacting with each other, walking in groups along ridgelines that run along the walls or fighting with each other or even preparing to mate—but not the horses. They stand quietly, some in small groups, some solitary. In one section, horses looking just like the gang of rogue "teenage" males Jason Ransom and I had seen in Wyoming stand in a row, side by side. One has his head down and mouth open, as if just looking up from his grazing. His wide eyes and pricked ears portray alarm. Another is just beginning to pay attention to some distant danger. Another is poised to flee. Below the horses, two rhinoceroses lock horns. Is this what frightened the four horses? Or were they startled by the prowling lions depicted nearby? Researchers have studied this particular panel in great depth and have discovered the order in which the animals were drawn. The grazing horse was added last, like the teller of the tale who just happened to be present.

In another scene, two horses face each other. Their necks are arched. Perhaps they are trying to stare each other down, like the two Pryor Mountain stallions Ransom and I saw. Elsewhere, a lone horse peeks timidly around a corner. His hindquarters are not drawn, so that he seems to be emerging warily from some crevice, perhaps trying to find out if it's safe to come out. Is the valley scene, filled with chaos, frightening him?

Lots of theories purport to explain the genesis of Chauvet. The one thing that we do know is that the artist, or artists, understood animal behavior. Craig Packer, an expert on African lions and one of the rare scientists who has been allowed inside the cave, came away impressed not only with the art's quality, but with the depth of the artist's knowledge. In the scene of two mating lions, Packer told me, both the male and the female lions behave in Chauvet just as he had seen modern lions behave in Africa.

What should we read into Chauvet? Several French authors suggest that Chauvet was the creation of shamans who hoped to evoke the animals' spirits. Others suggest that the complete Chauvet frieze running through the narrow cave can be understood almost like a modern-day graphic novel. In this view, the artist was telling a story panel by panel. One recent idea suggests that the animals would appear to be alive when tallow-lamp light flickered over their images, creating the illusion of movement.

Chauvet is unique for its grandeur and its individuality of expression. But from the western coast of Spain all the way to Russia's Ural Mountains, images of horses abound. You could spend a year traveling from site to site and never see them all. Yet despite these vast distances in both space and time—it's important to remember that the phenomenon of Pleistocene art extends over a period of more than twenty thousand years—there are a few universals. The images are usually peaceful. Violence is rare. A few images have been interpreted by modern researchers to show animals with hunting spears in them, but such depictions are unusual. Humans almost never appear; when they do, only a few are shown injured. And, oddly, landscapes—trees, flowers, rivers, cliffs—never *ever* appear.

Among the animals routinely depicted, horses seem to hold a special place. They are almost angelic. Stallions during the Pleistocene must have battled the way Ransom and I saw stallions battle in the Pryor Mountains, but the art never shows these confrontations. It never shows mares biting each other, never shows foals being attacked by wolves or lions. Other animals are involved in violence, but, until the very end of the Pleistocene, the horses always seem to be peace-loving.

There are some differences in style, though. Whereas the exquisite

Chauvet horses were serious animals, the horses of Lascaux, created about seventeen thousand years ago, are merry little things, delightfully capricious. The renowned author and archaeologist Paul Bahn sees Pleistocene art as "an art of tenderness," and that is certainly true in the magnificent cave of Lascaux. Lascaux is all about horses in motion. Horses are present in abundance here, accompanied by aurochs, bison, bovines, stags, ibex, and even some large cats.

In Lascaux, the horses' motto seems to be "Don't worry. Be happy." The attitude makes sense: temperatures were warming, but the extensive flat grasslands of the European littoral were still filled with wildlife. It was a great time to be alive. And so Lascaux is filled with joyfulness, a bit of craziness, and plenty of variety. On its walls are dark horses and light horses, pintos and bays, a horse that some people think looks "Chinese" (because it seems to resemble horses in Chinese art) and horses that some say look like tarpans, a now-extinct type of European horse. These horses wander over the cave walls in small bands, just as they wander today.

They are positively prancing in their excitement. Lascaux is full of happy enigmas, like a cow that seems to be leaping over other animals. There are well-fed horses with roly-poly bellies, satisfied mares accompanied by gamboling foals. If that isn't chaos enough, there's an upside-down horse, perhaps rolling in the grass, surrounded by other horses. Were all these horses and other animals portrayed as a group? Or did individual artists just create individual horses? At Chauvet, it's possible to see the art as one huge piece of work, but Lascaux doesn't have the same sense of unity.

Elsewhere, the horses are quite different. On a warm May morning, shortly after I visited Isabelle Castenet, I walked up a stream valley in the Basque country to the mouth of Ekain, a painted cave that's only about fourteen thousand to twelve thousand years old. The walk was an easy one, along an unpaved well-worn cart path that may have been the main route through this valley since the days of the Neanderthals. I had come to see a lesser-known panorama of horses.

Along the way I met a local man to whom I said, "*Buenos días*."

He quickly corrected me. In the Basque country, the proper "Good morning" is said in the Basque language, a language unlike any other European language and which may, in fact, have roots that stretch back to the Paleolithic era. I had a frustratingly difficult time wrapping my tongue around the words he repeated to me.

I reached the mouth of the cave and turned around. Spread out below me was excellent horse territory, rich with flower-filled meadows, as in the Wyoming high peaks. Two small streams met just below where I stood. They had carved a slight neck of land that was covered with greenery. Beyond was an even larger, richer valley. Above me rose some steep hills. Pleistocene horses with thick legs would have had little trouble negotiating these slopes.

In the Basque country today live free-roaming Pottok horses. Believed locally to descend directly from Pleistocene stock, these living horses resemble the graceful horses painted on local cave walls. (Studies performed by the geneticist Gus Cothran show that Pottoks and the mustachioed Garranos are indeed related.)

Watching modern Pottok horses can be a noisy experience. The lead mare is often collared and belled, like cows in the Alps. Like their Garrano cousins, the Basque horses allow people to get close—but only so close. They definitely have a critical distance closer than which they prefer humans not come. In the past Pottoks were captured and used for various beast-of-burden tasks such as carrying heavy loads over the mountain passes. In coal mines the little horses were used as pit ponies, condemned to live most of their lives underground, pulling coal-filled carts in the dark. Whenever the poor animals were allowed out into the sunshine—*if* they ever were—they were blind for days until their eyes could adjust. Many became permanently blind. The early novels of D. H. Lawrence, particularly *Sons and Lovers*, described the lives of these unfortunates with great compassion.

In the modern era, thankfully, their lives are better. Said to have gentle natures, they are sometimes trained for riding, although this apparently doesn't always work out well. I spoke to one barn manager who let little children ride them, but only if an adult carefully led each pony by a head strap. "They're too stubborn," she told me. "They're mostly wild and allowed to roam, so they don't like to do as they are told."

Pottoks, now sometimes crossed with Arabians or other breeds, have been successful in the show ring, but the pure ponies, left alone on their mountains, occupy a kind of netherworld. They are flighty. I held out a handful of grass to one who had spent the afternoon carrying children on pony rides. He snatched it away, but, like a wild thing, when I tried to put my hand on his head, he was gone in an instant. Chances are that when his work that day was over, he would be turned back out to roam the hills at his leisure, making all of his own decisions, until called up for duty again.

Pottok horses are sometimes called "semi-feral," but the term is nebulous. "Feral" implies that the horses were once thoroughly domesticated and then escaped to the wild. This is not the case, as far as anyone knows, with either the Pottoks or the Garranos. Local people say these Basque horses have always roamed these hills freely, so would not have "escaped" in the conventional sense of the word. The local archaeologist Pedro Castaños believes that Pottok horse populations plummeted at the close of the Pleistocene, but that wild remnants found refuge in the Pyrenees and have been there ever since. A more accurate word to describe modern Pottoks might be "semi-managed."

Throughout my travels in the Basque country, I saw these Pottoks everywhere, and I wanted to know how the horses of Ekain cave compared to these living animals. Tourists cannot go inside the actual cave of Ekain—the drawings are too precious to risk—but an excellent replica of the cave and its art is only a five-minute walk from the mouth of the real cave. A tour guide takes a small number of people through the replica twice. The first tour is done in complete silence; the second is narrated.

The silent experience is sublime. At the cave entrance is a drawing of a horse head, done in charcoal. Walking deeper into the cave, all that's audible is the slow, soothing drip-drop of water falling into a small pool. On the walls of the central cave are groups of sensitively drawn horses. There are other animals, too—more than sixty in all—but horses are the most common. Some are sketched in great detail, with their coats filled in with red ocher. Some are pinto-like, with coat colors that are black and white, just like some modern-day Pottoks. Others are merely outlined. They look like calligraphy. Yet on other horses, you can see the fine hairs of the fetlocks. Still others have shoulder stripes. Whoever drew these horses took the time to draw them well.

These are stylized works, not at all like the Vogelherd carving, the joyous horses of Lascaux, or the anxious horses of Chauvet. The lines on the cave walls evoke the concept of "horse," but not specific horses themselves. These horses don't look ready to rise off the cave walls and come to life. They do not seem to represent specific living animals that the artist once saw. They don't seem about to prance or to rear if we wait long enough. They are not observing other animals and they are not part of a grand scene.

Instead, the Ekain horses are representatives, or ideals. They reminded me of some of the most stylized images of horses in traditional Japanese art. Using only a few clean lines, these early Iberian artists showed the essence of the horse. The images seem to *mean* something, and that meaning likely goes far beyond "This is a horse." Perhaps there was a group of people who adopted the horse as their totem, or maybe just as their mascot. Or perhaps the horse represented a certain characteristic—sociability, maybe—that people admired.

This symbolic horse turns up everywhere in Europe and Asia at this time. Deep inside Monte Castillo, in a cave called El Castillo, where a wall bends at a sharp angle like the outside corner of a house, a graceful reindeer and a horse are paired back-to-back, one on each side of the angle. I had seen many two-dimensional images of this strange pairing, but the three-dimensional reality is celestial. Had the guide allowed it, I would have stood there for quite a long time. The animals seem to be free of gravity. The artist would have had to work at a challenging angle to create this illusion, yet both animals look peaceful. The horse is static. His legs are not depicted in any kind of gait. He has not adopted a noble posture. Rather, his head and neck are relaxed. He is not a particular horse, but a concept: "horse." The reindeer, too, is relaxed, but he seems, from our modern vantage, to have a gleeful eye. The fine lines of both animals tell us that the artist was highly skilled and *could* have drawn animals more like those of Lascaux or even Chauvet, but chose not to. Instead, the horse and the reindeer are just there, floating back to back, large and unmoving.

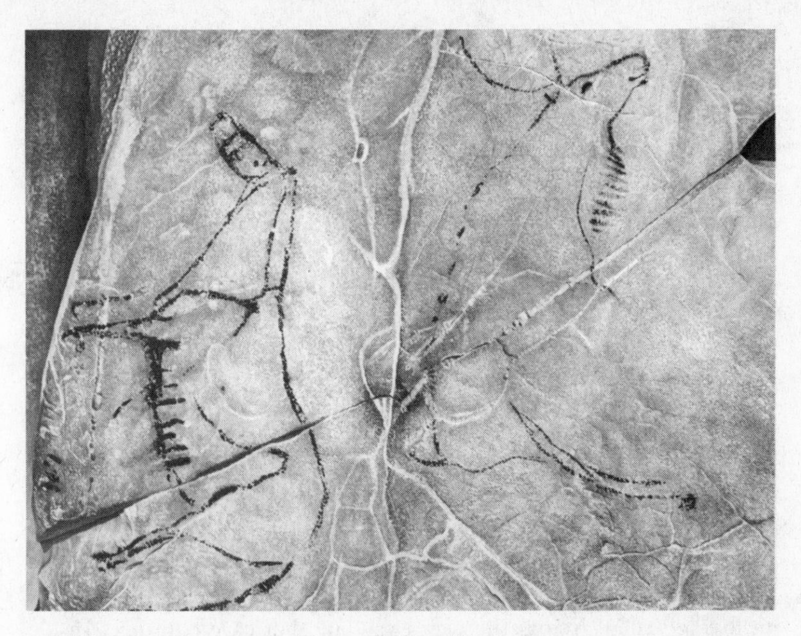

The ethereal reindeer and horse of El Castillo Cave (Courtesy of Paul Bahn)

We can tell from this art that, although horse population numbers were dropping, horses continued to be culturally important. Near Duruthy, France, on the eastern side of the Pyrenees, researchers have discovered what some say is a "shrine" to the horse. The Duruthy shrine—or more conservatively called an "ode to the horse" (as "shrine" implies religious worship)—was created at roughly the same time as Ekain's horse art and as El Castillo's horse-and-reindeer pairing. Duruthy researchers have discovered thousands of Paleolithic tools and artifacts at this site, including a horse carved out of sandstone. His muscular shoulders and head, with glaring eye and ears flattened against his head, make him appear to be lunging forward in a threatening posture, like the snaking stallion in the Wyoming mountains.

Evidence of this late-Paleolithic idolization of the horse can be found in the Russian cave of Kapova in the Ural Mountains, six thousand miles to the east of the Pyrenees. Roughly fourteen thousand years old, Kapova is situated in what's now a nature preserve located north of the Black and Caspian Seas. The cave shows well-nourished horses drawn in red ocher. They have thick necks, plump rumps, and delicate heads. The Kapova horses do not interact with other animals. There is no large scene or story line, as at Chauvet. Instead, the animals are friezelike. One horse trots in front of other animals, looking straight ahead. We don't know if he was meant to be shown as the leader of the animals or if the artist just set out to sketch various individuals in no special order. Elsewhere in Kapova there's another line of animals. Here, the horse jogs in the middle of the line, between a mammoth and a woolly rhinoceros.

Another frieze of horses from this time, called the Magdalenian, was done in one of my favorite rock overhangs in France—the Cap Blanc site, located not far from the town of Les Eyzies. This site is still open to the public, and visiting it is a simple matter of buying a ticket. Tours are given at certain specified times, and since I had an hour before I was scheduled to visit, I decided to hike. I descended a narrow path leading from the overhang to a stream below and found myself in an idyllic valley. Across the way I saw another cave entrance. Inside was another beautifully carved horse head. Had the people living at Cap Blanc fifteen thousand years ago decided to go visiting, it would have taken them only minutes to walk down their hill, cross the stream, and enter this other cave.

Why were so many horses created by artists in this area? Were the people who lived at both of these sites members of one extended family who had made the horse their family crest? During the Pleistocene, the relationship between horses and humans may well have been similar to the relationship between Inuit and whales, or between early North Americans and bison. On the island of Great Britain, in Gough's Cave, sometimes called the "Cave of the Horse Hunters," researchers have found many horse bones cast off by prehistoric people. The manner in which some of those horse bones are broken suggests that people were eating the marrow inside the bones. The people of Gough's Cave also carefully removed tendons from the horses' lower legs, possibly to be used as some kind of binding material.

Given this relationship, it's logical to assume that the art of the horse was somehow connected to human survival. Perhaps depicting the horse was a way of thanking the horse for the gifts he provided.

On the other hand, some French researchers have suggested that the horse, commonly paired with the bull, was a sexual symbol. One researcher asserted that the bull was male and the horse was female. Another Frenchman concluded the opposite: that the horse was male and the bull female. When I asked Lawrence Straus about the symbolism of the horse, he answered: "Who knows?"

What we do know is that the horses depicted in art from about thirty-five thousand years ago to just about ten thousand years ago were created by many different hands, and that the techniques of the artists varied greatly from age to age and from location to location. The creations range from fabulous depictions, those of Chauvet and Lascaux, to the mundane, as at Les Roches. The paleontologist Dale Guthrie, an avid hunter, spent years studying European Ice Age sites and ultimately decided that at least some of the art was mere "graffiti" drawn by teenage boys who were thinking about the horses they intended to hunt. For Guthrie, the human impulse to create art is a natural form of play behavior in which the boys, preparing to become hunters, indulged.

The anthropologist Genevieve von Petzinger cautioned me against thinking of horses in Pleistocene art as representing any one thing. After all, the horses of Lascaux are half the age of the horses of Chauvet. "In all likelihood there is no grand unified theory, no one single meaning," she told me in a phone conversation. "What we're dealing with is thirty thousand years of history and multiple cultures, each of whom may have had their own reason for using the horse, and even within those individual cultures, they may have had different reasons for depicting that particular animal. It's always a big problem. People want to know what the art *is*. Well, it's probably lots of different things to lots of different people."

The horses reflect the ongoing nature of the horse-human partnership that I had first thought about standing with Phyllis Preator atop Polecat Bench. In one form or another, the horses had always been with us, and by the Pleistocene era, *Homo sapiens* had become sophisticated enough to be able to express their pleasure at this companionship. In

Horse, the Canadian author J. Edward Chamberlin summed up the Ice Age artists quite simply as "those who praise horses."

Nevertheless, while they were praising horses, they were also eating them in great numbers. While there is no clear evidence of North American early people hunting horses, in the Old World record there are a multitude of undisputed sites. Indeed, the history of the predator-prey human-horse relationship stretches all the way back to the days when *A. afarensis* and the *Hipparion* mare and foal walked the Laetoli plain. Not far from Laetoli archaeologists have found very primitive stones that may have been used as butchering tools in close association with broken-up horse bones dating to about 3.4 million years ago. Bouri, a 2.5-million-year-old African site, clearly shows tools fashioned by *Homo erectus*, ancestor to *Homo sapiens*, near butchered horse bones. This improved ability to gather meat is widely believed to be an important reason why the *Homo* lineage was able to evolve ever-bigger energy-greedy brains. A vegetarian diet of fruits and grains gleaned from nature would not have provided enough calories to nourish larger brains, so that the ability to find meat, the ability to manipulate objects such as stone weapons, and the evolution of large brains probably went hand in hand.

There is also plenty of evidence of humans eating horses as they migrated out of Africa. At Dmanisi, at a site located between the Black and the Caspian Seas, on a prime perch occupied by members of our lineage for possibly two million years, butchered horse bones are prevalent. In southern China, horse bones have been found in association with *Homo erectus* at a site dating to 1.7 million years ago. In Britain, at a site called Boxgrove, half-million-year-old bones of a more recent human relative, *Homo heidelbergensis*, are found near horse bones. A horse scapula there has a hole in its center, which archaeologists say may have been made by a hunting weapon. Around one horse carcass researchers have found flakes of stone, so that they believe these early people were sharpening up their stones even as they were butchering the carcass. "You can almost imagine the stone waste materials piling up between the legs of the flint workers as the hand axes took shape," wrote the archaeologist Douglas Price in *Europe Before Rome*.

In Germany, not far from where the Vogelherd statue was discovered, a spectacular cache of wooden spears said to be three hundred thousand years old—one hundred thousand years before *Homo sapiens* evolved in Africa—has been found at what German researchers call a "wild horse hunting camp." Probably made by *Homo heidelbergensis*, these are the oldest known hunting spears found anywhere in the world. Of the thousands of animal bones found at this site, Schöningen, almost 90 percent are bones of horses. In southern Spain, at a site called Abric Romaní, researchers have discovered bones of many different animals, including horses, left by Neanderthals.

By the time the Vogelherd horse was carved, some *Homo sapiens* were hunting horses in a highly organized fashion. *Homo neanderthalensis* in Spain hunted horses one at a time, but by thirty-five thousand years ago, *Homo sapiens* knew how to hunt them band by band. The archaeologist John Hoffecker theorizes that by then *Homo sapiens* had developed a "super-brain" ability to pool their brainpower and communicate and cooperate with each other that went far beyond anything the Neanderthals could accomplish. The art itself may have been part of this "super-brain" culture, as it allowed people to leave communications that stayed on the cave walls long after the artists had left. As evidence of his super-brain theory, he cites locations from Russia to France where people took down horse bands by working together to ambush horses traveling along commonly used animal trails.

Hoffecker also studied a site on a floodplain of the Dnieper River in Ukraine where he found indirect evidence that humans thirty-two thousand years ago may even have used some sort of constructed trap into which people could drive the horses in order to more easily hunt and butcher them. We know this technique was used elsewhere: at another site, Kostenki, in Russia, Hoffecker has found evidence that people hunted large numbers of horses by driving them up into box canyons.

There are also sites in France showing that *Homo sapiens* hunted horses in large numbers. At Solutré, nineteenth-century archaeologists found thousands upon thousands of ancient horse bones. One fanciful author suggested that people drove the horses up the steep mountain and then ran them over a cliff, but this is unlikely. New findings suggest that hunters lay in ambush along a commonly used animal trail that ran around the base of the cliff. When an unsuspecting lead mare brought her band around a bend in the path, the hunters pounced.

The number of these horse kill sites diminishes as the ice ages ended, suggesting that the numbers of horses were also diminishing. But they never entirely disappeared from the archaeological record, and the custom of holding feasts in order to share horse meat lasted well into the modern era.

We think of the Pleistocene as an era unto itself, as a time when people were so different from the way we are today that we might barely recognize them as human. But in fact they were quite like us, minus the conveniences of electricity and indoor plumbing. We can see that in the extensive records they left for us in their living spaces, and we can see that, during that time, something special was happening between horses and humans—something that went far beyond the typical predator-prey relationship.

What's particularly intriguing about the horse art left behind by these people is that many of these early artistic memes have lasted throughout time. We can see the same idealizations of the horse in Greek friezes, in Renaissance works, and in modern paintings. The same elegantly curved lines used to depict the Ekain horses show up today. Indeed, everywhere in Europe, the horse has remained iconic. The massive Uffington White Horse of England—almost four hundred feet long, carved into a chalk hillside about eighty miles west of London and maybe three thousand years old—doesn't look much different from the horses of Ekain. Greek sculpture from twenty-five hundred years ago shows a threatening horse head, with open mouth and laid-back ears, just as in the twelve-thousand-year-old Duruthy horse.

The arch of the neck found in the Vogelherd horse is still a popular meme, although its meaning has evolved. During the Middle Ages and through the nineteenth century, horses with powerful necks were often shown ridden by kings and princes, so that the horse reflected the power of the rider rather than his own. The seventeenth-century master artist Velázquez showed the Spanish king Philip III subduing a stallion (who represented the people) by forcing him into an obedient *levade.*

Then, in the modern era, the art completed a full circle. The horse became once again an innocent, as at Chauvet. Franz Marc's paintings emphasize the dreamlike nature of horses. In *Dream, Think, Speak*, the twentieth-century British artist Christopher Le Brun placed a white horse at the center of an ethereal canvas. Le Brun's horse is just there, floating in space, like the horse of El Castillo.

In what is to me the most moving modern "ode to the horse," at the Reina Sofía in Madrid, Pablo Picasso's *Guernica* is displayed. At the center of this stark black-and-white canvas, more than eleven feet high and twenty-five feet long, an agonized horse struggles hopelessly, trying to rise off the ground. His nostrils flare in terror. His mouth is wide open. His body is tangled and twisted. He is paired with the bull, as happened so often in Pleistocene art. This may not be coincidental, since Picasso was enamored with Ice Age art.

Guernica was occasioned by a real event, the world's first carpet bombing of a civilian village. On April 26, 1937, German and Italian bombers, at the behest of the Spanish dictator Francisco Franco, flew over the Basque town of Guernica, not far from where the Pottok horses roam today. The planes killed the town's animals, women, and children while the men were away working. Picasso began *Guernica* only days later.

When I visited, groups of somber people stood quietly.

I asked Sebastián Jurado Piqueras, a guide in the hall, to explain the horse.

"For me," he said, "the horse is the most important character here. His tongue is pointed because his scream of pain cuts like a knife. He shows a suffering of the people that is so profound that it cannot be expressed in words."

Picasso himself refused to explain the horse or the bull. He was as silent on that subject as the Pleistocene artists he so admired.

Once, when pressed heavily, he replied: "This bull is a bull, and this horse is a horse."

7

THE PARTNERSHIP

> . . . a man on a horse is spiritually as well as physically bigger than a man on foot.
>
> —JOHN STEINBECK, *The Red Pony*

In the fall of 2013, more than fifty free-roaming New Forest ponies in Britain's New Forest National Park died because of a climate anomaly in the warming Atlantic. That year's spring rains had been excessive, flooding large expanses of Europe. The summer had been hot. The forest's plentiful oak trees were fruitful. By mid-autumn, a mass of acorns covered the forest floor.

Horses love acorns.

In fact, when it comes to acorns, most horses just cannot help themselves. Sadly, what heroin is to people, acorns are to horses: an addiction that kills. The tannins and other toxins in acorns overwhelm the horse's kidneys, which then shut down. The horse becomes lethargic and dehydrated. He eventually dies. There is no cure.

Horses are immensely powerful animals, but in certain ways they are terribly, awfully fragile. People who partner with horses know that they must keep acorns out of paddocks and pastures and that they must not allow the horses to eat even the leaves of oak trees. Some owners with oak in their pastures fence the oak off, taking care to pick up fallen acorns. Others remove the trees entirely.

For more than a thousand years, the people of the New Forest have partnered up with the horses who roam there, protecting them by releasing pigs into the woodlands when the acorns begin to fall. Pigs love acorns as much as horses do—and they can eat all they want without getting sick. Local folk annually release several hundred pigs to clean up the acorns before the horses can get to them. This partnership in the New Forest has existed for at least a thousand years.

But in 2013 the pigs failed in their duty. The acorn crop was so large that even twice the normal number of pigs could not eat enough to keep the horses safe.

The horses ate the acorns and died.

Given the wide range of foods that wild horses eat—brush in the American West; beach grass along the southern Atlantic coast; tree bark at winter's end; sea peas on Sable Island—it seems odd that horses never evolved the ability to eat acorns. At the very least, you would think they would have evolved the instinct *not* to eat them. That never happened. There may be an explanation for this: For most of the horse's 56-million-year evolutionary history, oak trees and horses did not commonly encounter each other. For the most part, oaks did not grow where horses grazed, and if they did grow in horse territory, they were few in number.

That changed when the ice finally disappeared from most of Europe. Vast oak forests spread northward to cover the continent, just as they did in parts of North America. Oaks are unusual trees. Having originated long ago in what's now China, during the days of the supercontinent Pangaea, oaks were not immediately able to survive in many of the world's other ecosystems, but some species did spread, albeit rather slowly. At Ashfall, about 14 million years ago, at least twenty different horse species roamed in Nebraska. Not far from there, a fossilized oak tree has been found, indicating that oaks did exist in the area. But scientists know that those conditions did not continue. Recall that a layer of caliche—hardpan—has been found just on top of the layer containing the numerous horses (and the one oak tree), indicating a "significant drying event," in the words of Mike Voorhies. Following that drying, researchers found only five species of horses at Ashfall—and no oak trees.

But after the ice melted, oaks became prolific. They "positively sprinted" northward, in the words of Bill Streever, the biologist and author of *Cold*. More than six hundred oak species now grow worldwide. Like horses, they are tremendously versatile. Some produce acorns that are highly toxic, not just to horses, but to humans and many other mammals (except for pigs). Other oak species produce acorns that are less toxic, but that if eaten in large enough quantities can still poison an animal.

Oaks are not the only trees that are toxic to horses. Helicopter seeds from the tree *Acer pseudoplatanus*—a common species of tree—cause an often fatal myopathy, a wasting of the muscles. The American persimmon tree causes colic; the black locust tree causes cardiac arrhythmias; the chokecherry and black cherry bring about a fatal cyanide poisoning. Horses had quickly adapted to life on the open grasslands, but the sudden spread of forests may have been too much for them.

The spread of mixed oak forests was likely one of many ecological changes that helped bring about the extinction of horses in North America and the diminishment of horse numbers in Europe and parts of Asia. Even if a horse survives an acorn binge, the toxins can cause terrible damage. Pregnant mares will abort their foals, for example, limiting the number of offspring in the younger generation.

The British archaeologist Robin Bendrey is one of the leading scientists who agree that climate—and the subsequent ecosystem changes—rather than human hunting played a major role in the extinction of horses in North America and the near extinction of horses in Europe. "We're moving away from the blitzkrieg hypothesis—but it's become solidified in the literature and it's hard to shift. We're thinking that this was a much more nuanced thing, a question of animals expanding and contracting, of changing environments," Bendrey told me. He cautioned that the record in Europe is "sketchy" for the five thousand years following the end of the Pleistocene, so that it's difficult to say anything definitive.

It is true that the record is sketchy in Europe, but it's at least more thorough than the North American record. The available information points to a changing climate that shifted ecosystems, spreading thick forests over once-open plains. Europe and Asia changed in other ways as well. The land not covered in mixed oak forests was often instead covered by the encroaching sea. When the ice melted, sea levels rose. The levels never got as high* as they were during the days of the Eocene, when much of Europe was islands, but the sea did eat away substantially at the land.

To understand the magnitude of the shift, it's helpful to think about what Ice Age life was like for Eastern Hemisphere horses *before* the ice disappeared. Roughly seventeen thousand years ago, the Great Northern European Plain was a wide-open, extensive grassland paradise. In the millennia following the Last Glacial Maximum, dated to roughly twenty thousand years ago, when ice from the north had reached its greatest southerly extent, had he been so inclined, a horse could have easily wandered from the westernmost edge of the plain, where Galicia now is, all the way east through central Europe to the Ural Mountains. From there the horse could have wandered north of the Caspian Sea and on into the center of Asia. He could have loped over the rolling land of Mongolia to finally arrive in Siberia. On each leg of his journey he would have found good pasturage and a gentle topography. There would have been few rugged mountain ranges to deter him.

Had he not been inclined to travel to Asia, he might have opted for a northern route. The horse would have been able to travel quite far toward Scotland without having to swim. He could have walked from Germany to London and points north without ever seeing salt water.

The wide plain was also a land of opportunity for people, with fabulous hunting. Researchers suggest that the people living in refuges such as northern Spain and southwestern France may have traveled up there to hunt in the summertime. Other people may well have lived there year-round. Evidence of their presence is plentiful: fishermen have dredged up artifacts of Ice Age human activity, including plenty of mammal bones and horse bones, from the bottom of the North Sea.

For horses and humans, the slowly warming world was a genial place. We are both savanna animals—"hard-wired to be mobile," in the words of the archaeologist Barry Cunliffe—and in those days, there was lots of savanna to explore. But as the Pleistocene continued to wind down, the rising seas flooded the region. The flat expanse

* At least, they haven't yet.

connecting Germany to Britain contracted, so that the once-wide plain began to look more like an isthmus. Then, finally, the seawater rose high enough that Britain became an island. This was not a "Noah's Flood" event, not the kind of sudden inundation that happened in many parts of North America, but was instead something that happened slowly over thousands of years.

As the land disappeared, so too did the horses. Only fragments of their former habitat remained, inhabited by relict populations of the once-common animal. Many of the expansive estuaries that are such a beautiful part of western Iberia's coastline are really no more than the upriver remnants of wide river valleys that led to wide littoral plains on which grass grew. The horses who managed to survive the change in climate (paleontologists and archaeologists do find a few horse bones from this period, but not many) were all that was left of the once-ubiquitous horse bands.

Iberians believe that the modern ponies of the Basque country and of Galicia are descendants of those relict bands. The Sorraia horses found in Portugal may also descend from Pleistocene survivors, as might the New Forest ponies. Felipe Bárcena hypothesizes that the New Forest ponies are really descendants of Garranos transported to Britain by Galician ships thousands of years ago. We do know that at least a few horses remained on that island after the ice melted. Their eight-thousand-year-old unshod hoofprints recently appeared at a site called Formby Point, on Britain's western shoreline just a bit north of Liverpool. Buried long ago, they reappeared when modern tides washed away the ancient soil layers. What we don't know is whether the relict populations continued and provided the seed for the ponies of Britain, or whether, as Bárcena suggests, those modern ponies came later from elsewhere in Europe.

This is one of the most frustrating problems in the history of horses: Where, exactly, did the horses survive? We know that they did survive, or there wouldn't be horses in the world today. But because of the spottiness of the record in the millennia following the end of the Pleistocene, which populations of horses survived to contribute to today's equine genetics remains foggy, at best.

We do know that the encroaching forests presented many formidable obstacles in addition to acorns. The trees were difficult for the horses to navigate. Predators found many hiding places. Forage was limited. The horses' main defense, speed, was no longer effective; the trees slowed them down.

In any case, forests generally do not support large mammals. The arrival of the mixed oak forests coincided with the extinction of the mammoths. Reindeer headed north, never to return. Horses retreated to a few areas where forests could not take root. Sometimes other animals pushed back the forests, helping the horses. In modern Africa, for example, elephants are de facto land managers who keep forests from spreading by eating young trees.

In Europe, of course, there were no elephants, and the mammoths were gone. But there was another de facto land manager—*Homo sapiens.* After the ice melted, Pleistocene people turned from living under rock overhangs to living in small open-air settlements and, in doing so, changed the landscape. By the time the Holocene epoch, our own epoch, began 11,700 years ago, *Homo sapiens* were skilled at using fire in many ways. They improved their hunting opportunities by burning the land to encourage the growth of young shoots, which in turn lured grazers. They cut trees around their settlements, opening small clearings. When they abandoned these settlements, the horses could come in to graze. No one knows whether any of these horses were domesticated, but since there's no clear evidence that they *were*, scientists have left the question open. Unfortunately, that's sometimes interpreted in discussions by the general public as meaning that they weren't. Perhaps the best way to explain this is that science—which relies on evidence—has no clear finding on this matter.

In other places, forests just didn't thrive. In Galicia's extensive coastal heathlands, for example, the soil was too acidic, and the wind and salt spray stunted the growth of any trees that managed to put down roots. Here, another ecosystem would prevail—one that the remaining horses were able to adapt to.

One absolutely frigid June day, in the midst of the climate anomaly that encouraged the growth of too many New Forest acorns, Jason

Ransom and I and several Galician scientists, including Laura Lagos, visited such a place, located in the extreme northwestern corner of Spain. I'd thought of Spain as a "southern" country with a hot and dry climate, but Galicia's heathlands sit on a parallel with Nova Scotia.

In a four-wheel-drive vehicle we drove over muddy, deeply rutted roads up to the top of a three-thousand-foot-high coastal mountain. The wind, coming in off the North Atlantic, was blowing at gale force. We pushed open the doors of the vehicle, turned sideways to the wind, and made our way to the edge of an abyss.

Our guide, botanist Jaime Fagúndez, opened his arms to show us the valley below.

"*This*," he declared, "is called a hyperoceanic system."

The term "hyper" was appropriate. We looked out as far as the eye could see, which turned out not to be very far at all. There was too much fog, too much mist, too many rain clouds.

Ransom and I looked at each other.

"*This*," we said, "is Spain in June?"

I wished I had my January-in-New-England coat with me. Were it not for the salt in the air coming in off the Atlantic, the rain that pelted us would certainly have been snow. Ransom, who'd had an entirely different impression of where we were going, had only sandals with socks. This area reminded me of pictures I'd seen of Sable Island—windy, wet, forlorn. I thought of an old British Victorian romance novel I'd once read: a distraught heroine runs frantically across the heath, only to be found the next day, dead from the cold.

There were Garranos here, but it was going to take an awful lot of dedication to see them. Despite the wind, these mountains were shrouded in thick, heavy mist. Fog made the valleys below us invisible.

This was not an unusual situation, Fagúndez assured us.

"The rain in Spain," I said to Lagos, "does *not* stay mainly on the plain."

"No," she said. "It doesn't."

Then she said: "Up here on these heathlands, it's like this pretty much year-round."

I snorted. Even my dogs wouldn't go out in weather like this—yet the Garranos we managed to see were downright jolly. My own horses typically stood dejectedly in this kind of downpour, with hanging heads and rumps to the wind, waiting for the nightmare to go away and for the sun to come out. If shelter was available, they sought it out.

I mentioned this to Felipe Bárcena, Lagos's senior colleague, who spent several days traveling with us.

"Well," he said, "they're not horses. They're *Garranos.*" Bárcena so strongly believes in this difference that he has proposed that they be declared a separate subspecies: *Equus ferus atlanticus.*

We had seen several other populations of Garranos much farther south, where the weather was kinder. Temperatures were warm. The days, sunny. Yet those animals had seemed rather sad. They had found ridgelines where the wind kept the insects at bay and stood with heads hanging. I was happy—nothing like the hot sun, from my point of view—but the horses looked like worn-out plow horses. They barely moved.

Our two species differ in our opinions as to what makes a great weather day, but before I began researching this book, I hadn't thought about how our weather preferences had anything to do with evolution. It turns out that horses, having evolved for a considerable amount of time in northern regions, rather enjoy chilly temperatures. Many of us, on the other hand, with our two-hundred-thousand-year-old roots in Africa, prefer sun and warmth. We can survive in colder climates, but the necessary skills must be learned.

As Ransom and I huddled in our vehicle, trying to keep warm and dry, the Garranos living in this "hyperoceanic" world trotted around, seemingly invigorated. They wandered up and down the steep hillsides. They chomped with great satisfaction on the gorse. The stallions argued with each other. The colts and fillies played games. Their long, matted manes and tails dripped in rivulets, which traced small patterns in the mud in which they stood.

"Maybe," I mused, "Garranos actually *like* miserable, rainy conditions."

The thought made me sneeze.

Watching them, it occurred to me that the Sable Island horses, whom I'd always thought of as victimized by the whims of the North Atlantic Ocean, might find their island home pleasant. This just goes to show how flexible horses have become over the course of their 56-million-year evolutionary journey.

Ransom and I talked again about his theory that horses today occupy an anthropogenic niche. Still, while these Garranos may have enjoyed the nasty weather, they had plenty of other problems. Living in these heathlands is an equally hardy, healthy population of Iberian wolves who have also stuck it out over millennia. We rounded a corner and found completely cleaned skeletons of a mare and foal. A tiny, perfectly formed little hoof and lower leg, still covered in hide, lay in the mud. We wondered if the wolves had followed the mare, waiting for those few vulnerable moments when she had to lie down to give birth.

In much of the rest of Europe wolves are either extinct or limited in numbers,* but here in the rugged Galician highlands they prowl at will. They take an inordinately high number of Garrano foals. On our travels Lagos pointed out plenty of scars on the hindquarters of the babies, making it clear that the wolves aren't always successful. We didn't see any actual wolves—they're too smart for that—but we saw plenty of wolf tracks and many piles of wolf scat.

Galicians used to control the wolf numbers by using wolf traps and organizing wolf hunts. The earliest known written record of a wolf hunt dates back about a thousand years, but the custom itself is probably much, much older. Lagos showed us several wolf traps—pairs of high walls built of stone. Each wall typically was almost a kilometer long. They are angled to each other in the shape of a very wide V. At the bottom of this funnel is a deep pit. During community wolf drives—before the invention of firearms—whole villages used to assemble and drive the wolves out of their dens by making noise. Fleeing from the harassment, the wolves ran along the stone walls, unaware they were running into a trap until the walls narrowed in and it was too late to escape. Along the run, other people jumped up from hiding places and made more noise, harrying the wolves and driving them forward. While one person would hesitate to behave this way with a pack of wolves,

* Although they are being reintroduced in a few remote places.

the villagers found safety in numbers. Apparently, they were able to herd the wolves without too much risk of bodily harm. The desperate wolves, driven by the group of people, ran forward into the V and fell into a pit, where people killed them. The Galicians preserve the walls as a cultural memory.

Contending with wolves is challenging for the horses, but I suspect that gorse, in its own way, is equally contentious for most grazing animals. It's a rather useless plant. It does grow leaves, but almost as soon as the leaves appear, they transmogrify into those ultrasharp thorns. Even worse: land taken over by gorse is land that is difficult to rehabilitate. The more you burn gorse, the better it grows.

On the other hand, gorse may be one reason why Atlantic coastal ponies survived: by learning to eat this marginal food, horses could live on marginal land. It may well have been an early example of horses opting for Ransom's anthropogenic niche: by eating gorse, Garranos could live in an area rejected by humans.

Small, stocky ponies, often well under five feet tall at the withers, with large heads and sagging bellies, these odd little animals are likely one of the main foundation lines of our modern horse breeds, particularly of solid-boned breeds like huge draft horses and tall, sturdy Irish hunters. Yet the Garranos are not wimps, despite their small size. They and their Atlantic coastal kin, genetic tests suggest, provided the sturdiness of leg necessary for the heavier work humans would eventually ask horses to perform, like carrying knights with armor in the Middle Ages.

Nevertheless, they are despised. Poor things. They're kind of like the Rodney Dangerfields of the horse world. Local folk differentiate between Garranos, whom they also call *bestas*—beasts—and other horses, whom they call *caballos*. All honor goes to the *caballos*; the *bestas* do the slave labor, like donkeys.

Galicians have tried to "improve" the Garrano through breeding, but to no avail. Lagos told me about one such effort, when villagers bought a well-bred stallion of another breed, a "better" stallion, and put him out with some Garrano mares.

The first year this "better" stallion was with the Garrano mares, she

A mustachioed Garrano mare gets a haircut. (Greg Auger)

told me, "he was very strong and there were many foals." But during the following winter, the stallion nearly starved to death.

"All he did was eat, eat, eat, all winter long," Lagos explained.

The following spring he was too weak to breed and had to be removed.

"Not only that—all the foals from that stallion also died," Lagos added.

I thought for a while about the meaning of the word "better." "Better" according to whom? Perhaps this comes under the categories of Mother Nature Knows Best and Let Well Enough Alone. If Galician scientists are correct and these ponies descend from the stock that survived the melting of the ice, then they evolved over millennia to fit into this unique coastal system. Horses brought in from elsewhere may simply lack the necessary biological "equipment"—like mustaches, perhaps—to survive on gorse and live in a wet, nasty, windy climate.

We know that following the environmental chaos that ended the Pleistocene, horses in the Western Hemisphere disappeared, and in Europe and Asia only a few distinct types lived on into the Holocene. Some type of thick-legged Spanish horse, like the Garrano or the Pottok, provided some foundation stock. A finer-boned horse type, ancestor to today's Arabians and Thoroughbreds, survived in the interior southerly deserts of Asia. There may have been another horse type that survived in Mongolia, and a few other small pockets of ponies, like Siberia's Yakut horses.

But in most locations, horses were either absent or nearly so. Scientists know this because archaeological sites from around ten thousand or so years ago are relatively common across Asia and Europe. In those sites, there are many bones of prey animals, but the bones of horses are found only very rarely, in contrast to the plentiful bones found at Pleistocene sites. In Greece, where bones of *Hipparion* are easily found, archaeologists have cataloged many early-Holocene bones of pigs—but not of horses, implying that horse populations had plummeted.

And yet, today horses live on all continents except Antarctica. Tens of millions of horses, the vast majority of them living in partnership with humans, inhabit the Australian outback, the plains of North America and Mongolia, the hills and pastures of modern Europe. Despite their susceptibility to insect-borne African diseases, they are now even common on that continent, because humans have learned how to inoculate the horses against illnesses such as sleeping sickness, carried by the tsetse fly.

We have gone out of our way to nurture horses, even ensuring that they can with our help live in areas where they could not survive on their own. We seem to just want them around, even though they are no longer necessary. We don't have to use them for transportation or farming, and yet they are still present on farms and ranches, and they still pull carriages, even if just for tourists. In Amsterdam, I saw a team of draft horses pull a wagon full of beer on a daily basis down one of the city's main streets. Every day, the traffic slowed to make room for the team. When the horses passed, people stood stock-still, entranced by their size and power. In Merrimack, New Hampshire, two-thousand-pound white-faced Clydesdale horses (with their feathery feet, reminiscent of the little Garranos) are kept for public relations reasons by a beer company.

Occurring in only ten thousand years, this is a phenomenal range expansion. It happened, geologically speaking, in an impressively short period of time. For this success, horses can thank humans, who have cared for, and even fussed over, equines in many different ways. I wonder if this is one way in which humans are truly different from other animals: We reach out to other species. And we respond strongly to animals who respond to us. As long as the animal isn't dangerous, we may even stand quietly, just watching, hoping that the wild thing we are spying on will come over to say hello. And it turns out that horses will sometimes do just that.

The closer they have allowed us to come, the stronger our partnership with them. We have bred them in large numbers, encouraging their proliferation, and we have transported them in ships from the Old World to the New, returning them to the continent on which they evolved. We have pampered them by feeding them rich grain, by ensuring their access to water, by driving away predators like wolves, and by sheltering them from the wind, rain, and snow. We have brushed their coats, looked after their hoofs, and even provided dental care when necessary.

If we have made ourselves useful to the horse, this is because the horse has made himself amenable to us. Sensitive, responsive, cooperative (mostly), comfortable to ride, and able to forage with limited human help, horses, as soon as they were domesticated, became the foundation on which human civilization was built. Researchers who study the earliest days of farming and animal domestication often compare the domestication of horses with the domestication of cattle and sheep, but there's more to the human affinity for horses than mere productivity. Horses, like dogs, are our *companions*. Because horses can form strong bonds, they may stay loyal to specific humans throughout the whole of their lives. And we may stay loyal to them: I often dream about horses I've spent time with many years ago.

Pleistocene art implies to me that this kind of mutuality may have existed for tens of thousands of years, but once we learned to keep horses in paddocks and learned to ride, the relationship became formalized. It's no exaggeration to say that the invention of riding was to early civilization what the invention of computers has been to us: a genuine, world-shaking Revolution with a capital *R*. Until riding began, travel by boat was the only convenient form of long-distance travel. We were, essentially, prisoners of the waterways.

But once people learned to ride, a new kind of ocean—an ocean of grassland—could be navigated, with only a little physical effort on our part. Horses offered people, for the first time, the experience of freedom. The interiors of Asia, North America, and South America were no longer barriers but beacons. Undulations of grass, daunting to a human on foot, become tantalizing temptations when we are mounted: Come hither, the grass whispers. See what's over this next rise.

When I read about the deaths of the New Forest ponies, and about the ancient custom of humans protecting them from eating too many acorns, I began wondering about the origins of this protective behavior and, indeed, of riding itself. Which step came first? When and how did riding begin? The conventional scientific view suggests that riding, along with the domestication of the horse, first occurred in central Asia. (It's important to separate the two phenomena. Riding is one event, archaeologically speaking, and horse domestication a different event. It's quite possible, for example, that people began riding horses long before they domesticated them.) We don't have an earliest date for riding, but we do have an earliest known date for domestication. Researchers investigating an archaeological site called Botai in Kazakhstan found what appear to be small "corrals" in which mares were milked. Bits of broken pottery recovered from these enclosures have revealed animal fats unique to horse milk, confirming the theory that mares were domesticated at this site. The site dates from about 5,500 years ago and is widely known as the first definitive evidence of people keeping horses. However, there is no evidence at Botai of riding.

Absence of evidence, though, is not evidence of absence. Horses could have been ridden or kept for milk long before Botai existed, but no one has found any definitive proof of this. Given that cattle and sheep were domesticated at least ten thousand years ago, and given that horses are better able than cattle to survive on cold northern plains (just like Whisper, they can use their hoofs to break through ice to drink or to scrape away snow to find grass; cattle cannot), it makes sense that horses would have been domesticated long before Botai.

It also makes sense that the horses would have been ridden. Riding does not *require* accoutrements like saddles and bridles. All you need to do is stay on the horse's back. If you can survive one or two wild rides, you're on your way. In any case, riding is part of our natural-history heritage. In Africa, I have often seen baboon infants riding on the backs of willing adults, reminding me of Huxley's funny illustration of Eohomo riding Eohippus. On the other hand, domestication of horses would have been a bit more complicated if by "domestication" you mean the act of controlling breeding in the modern sense of the phrase—leading the stallion to the mare.

The conventional archaeological belief that riding and domestication began in the east and spread to Europe annoys many of the

Galician scientists I met. They believe that both riding and domestication of horses were independently invented in their own region. Riding, they believe, was a natural stage in the progression of the horse-human partnership from the hunting of horses for meat to the routine use of horses in village life. Unfortunately, they can't point to any firm direct evidence, other than DNA reports of the genes of Spanish horses present in many modern horse breeds.

There does exist, though, some interesting post-Pleistocene art that clearly shows horseback riders. Their feet almost reach the ground, so we know that their mounts were quite small, probably about the size of the Garranos I had seen. This art, however, is not found in caves. The depictions are out in the open air, and consist of carvings made into rocks and onto the sides of cliffs. There are a lot of examples all over Europe of open-air rock etchings, but most are strange symbols that scientists cannot interpret. There are circles within circles, cross-hatched patterns that look like checkerboards, or simple X marks and sets of parallel lines. Some researchers theorize that these symbols may be very early precursors to writing—not writing itself, but a form of signposting or branding, as in "Kilroy was here" or "This land belongs to these people."

These geometric signs are common, but a second style of rock carving I saw with Lagos and Ransom is extremely rare, existing in Galicia but not elsewhere. These are certain kinds of petroglyphs that feature horses. This post-Pleistocene horse art, however, is quite unlike Pleistocene horse art. There is certainly nothing like the majestic Vogelherd horse. Pleistocene art focuses on the elegance of the natural horse, but this Holocene Galician art shows the horse as a tool of humans. People are now at the center of the action; horses are now animals to be subdued and used. They are objects of manipulation. The horse as wary watcher, as playful sprite, or as a member of a glorious gathering of large mammals had become a thing of the past. Replacing the Pleistocene's "art of tenderness" is an art that extols elitism. Those possessing riding horses are now in charge. As the British archaeologist Richard Bradley has observed, horses at this time symbolize nature subdued by humanity.

Ransom, Lagos, and I spent several days visiting these open-air sites. One scene shows a rider sitting comfortably in the middle of the horse's back. Researchers used to believe that people did not learn the correct seat on a horse's back until only a few thousand years ago (on Greek urns, riders are shown sitting too far back to be well balanced), but these Galician riders look quite secure. Their horses are not saddled, suggesting that saddles had not yet been invented. In one scene, the rider has reins in his hands that fly through the air over the horse's neck and head and connect to the horse's muzzle. It's unclear whether the rider is using a bit or is only looping a rope over the horse's nose.

Another scene shows what may be the earliest known depiction of a horse roundup. This is my personal favorite, since the same tradition continues in Galicia today. A rider accompanied by dogs is driving a number of horses through a barrier, maybe a stone wall like those constructed to trap wolves. Some horses have already been pushed through the gate in the wall. A few horses hang in the background. The rider is again shown with reins flying. One arm is raised, wielding a stick.

"That," Lagos said to me, "is the stick of power."

I looked and realized that I had seen something similar in the Prado museum in Madrid, in the previously mentioned portrait of a king riding a horse, painted by Velázquez only a few hundred years ago. The theme of horse, rider, and "stick of power" weaponry would remain a symbol for the elite well into the modern era.

I was enthralled by the idea that I might have been looking at the world's first depiction of a roundup, but when I asked how old the art was, I learned that, sadly, these works couldn't be firmly dated. Some researchers suggest that they may be as much as five thousand years old, giving credence to the Galician claim that they rode and domesticated horses quite early, regardless of what was happening elsewhere in the world. Others suggest a later date of three thousand or so years ago, by which time riding was quite common in many locations in Europe and Asia. This art, Galicians believe, provides evidence that riding horses began in Galicia. To show me how that could have happened, they took me to a modern Garrano roundup, one that closely resembled the roundup depicted in the stone etching from thousands of years ago.

When he drew Eohomo clinging merrily to the withers of Eohippus, Thomas Huxley was joking, of course, but it was a joke with a certain small portion of biological truth: riding may be instinctive. Many of my equestrian friends think this is true. We assume that people just naturally jumped on the backs of horses. Many archaeologists, however, disagree. To them, riding horses for the first time would have been a complicated cultural innovation.

Galicians disagree. And so Ransom and I were taken to a *rapa das bestas*, a "shearing of the beasts." They wanted me to see how easy it would have been for early people to capture, pen, and ride horses well before those horses were "domesticated" in the conventional sense of the word. Garranos roam at will over the Galician hillsides, but local people routinely harvest their manes and tails to make items like artists' paintbrushes and bows for stringed instruments.

This is done by rounding up the horses and driving them into a stone corral—just as they had done with wolves, and just as I'd seen depicted in the Galician rock art. Early one Sunday morning, Ransom and I stood with Bárcena on a hillside and watched the scene unfold below us. Horns sounded all over the wide valley. The shouts of men came from many directions, urging the ponies out of their hiding places. Whoops and cries filled the air. The Garranos came jogging along, band by band, keeping just ahead of the noisy people. This was a true community endeavor. People driving on the road above the valley stopped their cars, got out, and yelled, contributing to the effort, and then drove on.

The Garranos were not pushed into a wild run, but were moved along steadily. Every once in a while a stallion galloped for a few strides—often just to chase another stallion—but not for long. No one wanted any injured animals. The flow of horses gathered strength as more and more small bands were herded into the main group. The parade grew larger and larger. When they arrived at the corral, a large arena with high stone walls, the people stopped moving the horses forward. The Garranos responded by eating some gorse.

Eventually, people pressed the horses through a gate into the corral. Inside the corral, panic ensued. Horses like their personal space, and in these tight quarters they were too close to each other. There

Villagers round up the Garranos. (Greg Auger)

was a great deal of complaining, of neighing and whinnying and rearing. A few horses kicked out at other horses with flying hind legs.

Into this chaos waded pairs and trios of men carrying long sticks. At the end of each stick, a single length of rope was attached. The rope had been cleverly tied to create a series of slip-rope loops that could be dropped over a horse's poll and nose to form a headstall. The makeshift halter tightened when a horse pulled, putting pressure on his poll and on his nose. When the horse stopped pulling, the halter loosened. Most of the horses caught this way were easily subdued. Looping one single strand so that it could provide substantial control over a stallion was something I'd never seen before.

Yet it was a very simple thing to do. While there's no proof that Pleistocene people knew how to do this, they were certainly technologically advanced enough to have figured it out. It wasn't easy to drop the looped rope over the horse's head. It was much more difficult than using a lasso. Several men helped the man with the rope, often by keeping the other horses away.

First the men caught the stallions, whom they led out of the corral and released. These treasured stallions were allowed to keep their long manes and luxurious tails. They galloped a few steps in appreciation of their freedom but then hung around waiting for the mares.

Next the foals were caught. Some were released to return to the wild as well. These, too, stayed near. Other foals were led to a second corral, where buyers would look them over. One goal of this culling was to select the colts who would become future breeding stallions. A few archaeologists have written that domesticating the horse would have been difficult because controlling a breeding stallion is a dangerous task. They are correct in terms of modern stallion management, which often involves handlers leading a stallion directly to a mare. But there's a simpler way to control breeding: eliminate the stallions you don't want. Through this process of elimination, you allow the chosen stallion to sire the next generation without competition. At the *rapa das bestas*, undesirable colts were sold on, and the colts chosen to sire the next generation went free.

Next, the men caught the mares and brought them to the shearing area. One man held a mare by her head, using the rope halter. A second grasped the mare's tail, standing to the side to avoid being kicked. A third cut the mare's mane and tail. Once sheared, the mares ran off to their band members.

Events like these may have been where people learned to ride. In some of the Galician roundups, a team of men wrestle a Garrano to the ground and then clip the hair. After that, a man slides onto the animal's back. Another man grabs the horse's tail while a third grabs the forelock and the ear. Usually, the horse doesn't buck, but instead leaps forward and rears, pulling free of the two men holding the tail and forelock. The rider usually slips off after a few leaps, but we can imagine that a particularly athletic person could have stayed on and ridden until the horse tired. Tiring a horse out is an effective way of getting the animal to accept a rider. If the rider is heavy and the mount is small enough (some Garranos weigh eight hundred pounds or less), this could happen rather quickly.

This Galician event was a different style of horse management than anything Ransom or I had ever seen. These horses were not truly wild, in the sense that they were *never* handled by people. On the other hand, neither were they domesticated. Breeding is *influenced,* but not entirely controlled, by human decision-making. As Lagos had seen, mares can and do make their own choices.

What, then, is the Garrano, if not domesticated and not wild? They are sometimes, as I mentioned, called "semi-feral," but that term implies that the horses are escapees. Since there's no evidence that these animals were once fully domesticated, and since evidence points to the relationship between the Galicians and their Garranos as having roots that stretch far, far back into the mists of time, neither word really fits.

Watching the roundup, it struck me that the traditional division between "wild" and "domesticated" may be too rigid, and that for most of the history of the horse-human partnership, this informal style of keeping horses was probably the most common. After all, for most of our history, we didn't build elaborate barns for horses. And even fencing horses into anything larger than a paddock would have been difficult until barbed wire was invented in 1867.

Until very recently, livestock has generally been allowed to roam at will, accompanied by a herder. When I lived in Africa, I often saw small boys taking their goats and cattle out to graze in the morning and returning with them at night, when they were kept in a *boma*—a small enclosure made of bushes and twigs whose purpose was not to fence the animals in but to keep the lions out.

Horses, of course, didn't need to be defended from lions, since they could defend themselves. But since they are creatures of habit, keeping track of them may have been rather simple. The earliest ridden and domesticated horses were probably kept as modern Galicians keep their Garranos.

Ransom, Lagos, Bárcena, and I talked about this possibility as we walked away from the horse corral and headed toward the picnic tables, where, as at American rodeos, much food is shared and much beer consumed.

Bárcena bought us all wine and octopus and ice cream.

"How far back does this tradition go?" I asked him.

I got the stock answer: "For as long as anyone can remember."

With no artifacts and with only the Galician rock art as a written record, it's impossible to know. Domesticating sheep, goats, and cattle changed the skeletons of these animals in ways that can be detected by archaeologists, but that's not the case with horses. There is no substantive discernible difference between a domesticated horse and wild horses.

All of this suggests that the current debate over whether modern free-roaming horses should be called "wild" or "feral" isn't overly useful. Science is slowly accepting this viewpoint.

The British archaeologist Robin Bendrey talked to me about the changing view.

"It's always been argued that domestication is biological control—the management of breeding. But in the Near East, the domestication of other animals is beginning to be seen as a long, gradual process where you have a slow change in the relationship between humans and animals." In Bendrey's studies of goat domestication, he has found what he calls "a gradual change from the intensification of hunting to the goats associating themselves more with humans. This gradually evolved into early domestication."

I thought about his phrase "associating themselves."

How would that apply to horses—in my experience, animals who had always been forced into a relationship with humans? What in the world would horses get out of the deal? Is it possible that the domestication of horses was a two-way street—that horses might have *chosen* to associate with humans?

Bendrey cautions against thinking of domestication as a unidirectional, simple event. "It's often been viewed as—humans woke up one morning and decided to domesticate animals because they wanted to eat more," he said. "But there's two parties involved in this domestication event. It's not just a human-based decision." The partnership of humans and dogs or humans and horses may well have occurred long before the animals were "domesticated" in the conventional sense of the word, in part because the animals themselves benefited. For example, it's widely believed that northern people first domesticated reindeer by offering them human urine, which contains salts and other minerals the reindeer need in order to thrive during cold, dark winters.

As the reindeer became accustomed to this relationship, people became accustomed to following the reindeer.

Perhaps humans had something the horses wanted—handfuls of grain that they had gleaned from wild plants, or access to salt. Or maybe humans helped keep the wolves at bay.

Are there any modern-day examples of this kind of partnership that we can study to see how it might have been done?

As large flakes of Christmastime snow gently drifted down, I reached out to touch the muzzle of a "third-strike" mustang on Kris Kokal's farm in Greenfield, New Hampshire. New Hampshire is not a place most people associate with American mustangs, but on this farm twenty or so roamed the pastures that surround the farmhouse. The Kokal family specializes in rehabilitating mustangs that have been poorly cared for by other people. For very little money, the federal government allows people to buy mustangs removed from Western ranges. But since horses who have grown up independent of human supervision "think different" from horses raised around humans, these purchases do not always work out well. In that case, owners may not want to keep the animals, but they may have trouble finding a place to send them. Very few people want to take on the expense of feeding a horse that is difficult to ride or to work around. One place such horses can go is the Kokal farm.

My ungloved hand was still inches from the mustang's nose when the gelding turned his face away. I reached farther, making at least a bit of physical contact.

"Can I show you something?" Kokal asked kindly.

I nodded.

"Think 'deal' or 'no deal,'" he suggested, extending his hand toward the same horse. This time the horse stretched his head over and connected with Kris, allowing himself to be scratched.

"If he reaches out to connect, then you've got a deal," he explained. "If he turns away, then it's 'no deal.' You need to get that acknowledgment before you do anything more." To approach the horse after he's turned his head away is too impatient an act, he explained.

I could see his point. It was kind of like meeting someone halfway. There's only so much you can do and then the other person—or horse—has to take a step forward.

It often requires quite a while for one of these mustangs to choose to take that first step, given the trauma they've experienced at the hands of other humans. One small horse I saw at the Kokal ranch had been very much loved by a couple of kids who had had no money to buy food. Instead, they went every day to a local grocery store and took home the day-old leftover bread the store managers wanted to throw out. For three years, this horse lived on a diet of old bread, in a small paddock made of junk cars. He was emaciated, full of worms, infected with diseases, had hoof problems, and was very wary of people. A concerned neighbor bought the horse from the kids for one hundred dollars and called the Kokals, who agreed to take him. I saw him only a few weeks after he'd arrived. The horse was already more at ease, but he still had a long way to go.

The Kokals spend a lot of time with their horses. The horses are ridden, but most of the time the Kokals spend with the animals is on the ground, and some of the most important interactions happen while they and the horses are just milling around. Often when they hang out with the horses there's no particular goal in mind, no training that needs to happen. Throughout the day, someone will head out to the pastures and just stand there, not doing anything or insisting on any kind of goal-oriented behavior. The people and the horses are just there, in the pasture, together.

While we stood in the snow, Kris's mother, Stephanie Kokal, talked to me, and as she talked, she began scratching the dock of a mustang's tail. The horse responded by stretching his head out and nibbling contentedly at the withers of another horse. It's this kind of chain-reaction bonding that keeps a band together, and the horses seemed quite happy to let humans join up this way.

Stephanie scratched under the jaw of one of the most skittish of the animals. He dropped his head, half closed his eyes, and looked like he was about to fall asleep. It's not hard to imagine such an interaction happening thousands of years ago.

The Kokals never put the horses in stalls, although there are run-in sheds available in the winter months. At night, Stephanie says, the horses walk and walk, hardly ever standing still. Because of this, she believes that keeping horses in stalls affects their mental health.

Most of the Kokal horses are mustangs, but the family does keep two domestic horses. There are profound behavioral differences between the two groups. The domestic horses choose to shelter in the sheds when the weather is bad, but the range horses never do. They don't care for enclosed spaces. Additionally, the domestic horses keep their distance from the mustangs. When Kris walks into a pasture, the mustangs usually approach him. The two domestic horses are a lot less curious. Kris never brings anything like apples or carrots or other treats to the horses, and it's often someone else who brings the hay, so their interest in Kris is not based on food.

Their bond with Kris seems to be just that—a bond, like the bond formed between High Tail and Sam. Establishing yourself this way requires patience—not days and days, but months and months. Sometimes a considerable amount of time passes between when a horse first arrives at the farm and the time when Kris decides to ride the animal. Depending on the horse, riding may be one of his goals, but it's not the only goal, and it's never the primary goal.

The primary goal is getting to know the horse. "Watching wild horses is like sitting in church," Kris once told me. Just a time for peaceful contemplation and for unraveling life's mysteries. Once he's done that—watched the horse and gained some understanding of the horse's body language and temperament—then he may move on to the next step. "It's a beautiful dance," he said. "Once it starts . . . the horse really wants to be around us and everything can be done without forcing the horse."

To those of us who grew up in the sterner schools of equitation, this might seem a little far-fetched, but as Kris and I moved from pasture to pasture, the range horses all eventually trailed along with us. They didn't come over directly, following in a line. And they didn't come quickly. They just seemed to gravitate in his direction. The domestic horses, on the other hand, ignored us.

As the stranger in the pasture, I got a lot of attention. A four-year-old, still new to the rules of etiquette, began examining my nylon coat.

He was particularly infatuated with my coat's hood, which was rimmed with faux fur. After smelling the coat and determining that what looked like a coyote's pelt was actually nothing, he ran his supersensitive muzzle over my sleeve, playing with its feel. His lips touched the material so gently that they made a soft, musical noise as they swished back and forth. Whether he liked the sensation or not, I couldn't tell. But he was certainly engrossed in the exploration process.

Kris bought his first horse as a teenager. The family had grown up in a big city and knew nothing about horses, but he had always been fascinated with them. When the family moved from Florida to New Hampshire, he was finally allowed to have one. But the first time he tried to ride his new horse, the Appaloosa reared and dumped him on the ground.

"Send the horse back," he cried to his mother.

"I didn't send *you* back when *you* were trouble," Stephanie answered.

The horse stayed.

Left to his own devices, Kris had to figure things out by himself. Since his mother sent him out daily to his horse, and since he was reluctant to try to ride again, he just sat and watched. He wanted to know what the horse was about. He studied the horse's behavior. He made lists of what the horse did. He made lists of *when* the horse did what he did. Then he tried to figure out *why* the horse did what he did. His dad is an airline pilot—the family is big on checklists.

Then he got a little bolder. Stephanie got a call from the town's police chief, complaining that Kris and his brother, who had also bought a horse, were causing traffic problems. The boys still weren't "riding" the horses in the traditional sense of the term, but they were pulling stunts like standing on the horses' backs while the horses walked around in the field. People driving by were stopping their cars and getting out to watch.

If it isn't one thing, Stephanie thought to herself, then it's another.

In the twenty-first century, the Kokal family is unusual, but not unique. In fence-free Mongolia I saw horses who lived most of their lives in free-roaming bands that were tracked by humans, much as northern people used to track the reindeer herds, but not "herded" in the sense that sheep or cattle are herded.

When a traditional Mongolian rider wants to ride a particular horse, he drops a rope over the horse's neck and removes the horse from the free-roaming band. He uses the horse for a few days, then puts the horse back in his band. The horses adapt easily to the two different lifestyles. When called into service and saddled and bridled, the horse stands faithfully beside a fallen rider or willingly jumps up onto an open-bed pickup truck. I've seen them riding patiently on flatbeds without any side panels over deeply rutted roads without being tied and without jumping off. Yet when the same horses are released and sent back to the band, they easily resume their other life.

This is a system that's remained essentially unchanged for thousands of years (except for the pickup truck), and may well be how horses were kept at Botai fifty-five hundred years ago. The archaeologists David Anthony and Dorcas Brown, a husband-and-wife team, believe that people were riding and working with horses at least a thousand years earlier than the evidence at Botai suggests. They base their theory on evidence that around sixty-five hundred years ago, human history in the western steppe region and in eastern Europe underwent an abrupt change. People who had once lived in small, stable villages began to live in larger settlements and to develop power elites. Anthony and Brown attribute this change to the advent of riding and horse management. The change in lifestyle, they suggest, came about because mounted warriors became highly mobile, and the small villages were easily conquered by riders on horseback. As a result, a system of rule by elite males with horses replaced more peaceful and egalitarian endeavors, they propose.

Whether or not riding began as early as Anthony and Brown suggest, partnering up with horses certainly changed the lives of people who lived on the Eastern European and Asian steppes. In the centuries following Botai, horses began to be used to pull wagons that could be filled with a family's entire possessions, making it possible for whole families to live the mobile-home lifestyle. A steppe family could winter in a village with other people, then head for mountains to find solitude and good pasture once the snow melted.

"The key that opened the grasslands was the horse," Anthony writes in *The Horse, the Wheel, and Language: How Bronze-Age Riders from the Eurasian Steppes Shaped the Modern World*. In the course of their summer travels, he believes, early horse cultures passed on information and ideas. Horses thus created a new connectivity—absolutely essential if civilizations are to prosper.

In fact, there's no real reason why riding* could not have been part of the Pleistocene lifestyle. The archaeologist and author Paul Bahn has suggested that it might have been possible. I spoke about this with Bahn, who quickly cautioned that this is sheer speculation, as there is no clear-cut archaeological evidence of such a relationship—no bits or headstalls, no evidence of blankets being used as saddles, not even any evidence, as at Botai, of horses being routinely kept in enclosures.

But it would have been no big deal for a Pleistocene person to climb aboard the back of one of the little Ice Age horses and ride like the wind until the horse tired. This would explain the proliferation of horse art during the Pleistocene, particularly when, toward the end of the ice ages, evidence shows that horses were no longer common and people were no longer commonly hunting and eating them. Since Pleistocene art almost never depicts humans, the argument that the art of that age does not show horseback riders isn't meaningful.

The example of Kris Kokal shows us that even partnering up casually with a horse may not have been that difficult. All the first riders had to do was offer something to the horse that the horse wanted, maybe just companionship. Then, if you're patient and consistent, the horse will start to come to you. That's their nature. Neither is there any reason why horse domestication could not have occurred long before Botai. As I learned by watching the *rapa das bestas*, the initial stages of horse domestication could have been simple. We know that people who

* In a limited fashion.

hunted horses were skilled at this, so it seems reasonable to assume a smooth transition from hunting horses by corralling them to managing the horses by releasing those deemed valuable as breeding stock. Unfortunately, in the days before saddles and metal bits, we are left, scientifically speaking, without hard evidence, with little more than daydreams.

Learning to ride horses was a major advance for human civilization, but there's still the question of why horses held so much allure for us long ago, long before riding and domestication began. Why was the artist who carved the Vogelherd horse so inspired by horses, and why do we still feel that same attraction today? It is our own nature to want to form relationships with the animals around us, but what made the horse in particular so special? One answer might be that we have always been hypnotized by that phenomenon of evolution—the horse's massive eye. Scientists have even realized that the eye and its neurological connections can help us understand the inner workings of the horse's mind.

8

THE EYE OF THE HORSE

The Lion and the Horse were arguing as to which had the better vision. The Lion, on a dark night, could see a white pearl in milk—but the Horse could see a black pearl amidst coal. The judges decided in favor of the Horse.

—ARABIAN FABLE, quoted in G. L. Walls, *The Vertebrate Eye and Its Adaptive Radiation*

One morning years ago, the Canadian vision researcher Brian Timney noticed the horses in his field staring intently at something in the distance. Timney had looked in that same direction earlier and had seen nothing, but he looked again. Seven miles away hot air balloons of many different colors drifted languidly in the gentle blue sky. To Timney, the balloons were far from alarming—a festive sight on a summer day.

To the horses, though, the balloons were cause for concern—not because they were balloons, but because in the minds of the horses they were something else entirely, something horses had evolved over tens of millions of years to pay close attention to. The balloons were that nightmare of nightmares for horses: *unusual moving objects*. They were barely visible, but when Timney watched his horses watching the balloons, he saw that the horses had the kind of look all horses have when they are deciding whether to take another bite of grass or to make a run for safer pastures. After tens of millions of years of life on the open steppes, evolution has provided the horse with a very focused point of view: Better safe than sorry.

Timney has devoted most of his career to studying vision in humans, but the behavior of his horses intrigued him. Just exactly how well did the horses see the balloons? How did what *they* saw compare to what *he* saw? How far could a horse see? And what did they perceive when they looked? He saw the different colors of the balloons against the sky. What colors did they see?

I once performed my own informal equine visual experiments, also inspired by noticing unexpected behavior in horses. Unlike Timney, I wasn't interested in the detection of distant objects, but was instead curious about what horses can see right under their noses. Once, trying to get a horse to do some stretching in his box stall, I handed him chunks of bright orange carrots off to his sides, so he had to stretch his neck toward his rump, then down toward his feet and even between his feet.

When the carrot was in my hand, he did fine: this was a horse who was particularly fond of carrots. But when I accidentally dropped one of the bright orange chunks, he couldn't find it. Eventually I realized that the horse couldn't *see* the carrot lying there. To me, the carrot was obvious: an orange-red object lying atop straw bedding. It stood out like a sore thumb. Not to the horse. Although he seemed to be looking directly at the carrot, he didn't pick it up. Like most horse owners, I'd always figured that the horse's vision was pretty much like my vision, give or take a few small details. But this was weird. In the world *I* saw in that box stall, there was a carrot lying on the ground. In the world the *horse* saw, there was no carrot.

Just to be sure that I wasn't wrong about the carrot problem, I tried the same experiment outside. Maybe the stall was too dark. Again, I dropped the carrot on the ground under the horse's nose. Again, he couldn't see it. I directed his head down to an angle where, it seemed to me, he should be able to see it easily. No results. I pointed to the carrot with my finger. Again no results.

And just to be sure that he hadn't suddenly developed a lack of interest in carrots, I picked it up and offered it to him. A carrot in the hand was something he'd learned to expect. It was gone in an instant.

For thousands of years, horses and humans—two species with unique visual systems—have been partnering up to help each other see the world around them. We do this intuitively, and our ability to perceive what the horse perceives depends very much on plain old saddle time. The more you ride, the more you anticipate your horse's vision. If you don't learn this art, you'll spend a lot of time in the dirt. The horse does the same thing in reverse: a seasoned horse has learned that his rider will see certain kinds of things for him.

If well partnered, horse and rider are really two fundamental components of one living organism. This is the essence of the joy of riding. We often feel and respond to the horse's vision. For example, when a horse gets nervous because of some danger he thinks he sees around him, we settle him with our hands to let him know we see nothing there. If the partnership rests on solid ground, he will take our word for it. On the other hand, when we're out on the trail, we rely on our horses to see tiny details that we ourselves might miss. Consider the old Westerns: cowboys sitting around a campfire are alerted when their horses look toward the rustlers on a distant ridge.

Of course, we don't always know what sight will spook a horse. One sunny afternoon in Swaziland, for example, I rode through parkland with a genial local guide who willingly indulged me in riding up to a variety of wildlife. When I approached a hippo in a small pool of water, the hippo ignored me and my horse. The nearby crocodile did not. Assuming that my horse would not care for the croc, I firmed up my hands on the reins, waiting for a bolt. The horse, a Swaziland native, must have been used to this sort of behavior from foreigners, because he ignored my hands and continued to walk along, half-snoozing. Most likely the horse had seen the croc crawling through the tall grass long before I did. He just didn't perceive any cause for alarm.

Because of evolution, horses' eyes are highly attuned to notice even the smallest of movements in their sweeping field of vision. This is why my Swaziland horse probably knew long before I did about that croc rustling through the grass. On the other hand, scientists theorize that the horse discriminates only about ten thousand or so colors, while we discriminate *millions*. Moreover, the colors available to us are by no means all the colors that exist. There are tens of millions more hues, but we simply cannot detect them. Pigeons run rings around us in that regard, proving that there's much more to the world than our eyes can pick up.

Humans, horses, and all mammals enjoy the same basic visual biology—another result of our shared evolution. We have camera eyes, complete with cornea, lens, retina, and optic nerve. But because horses and humans have traveled separate evolutionary pathways for at least 56 million years, the horse eye and the human eye have also evolved some important differences. Take color, for example. Horses, like most placental mammals, have only two color-responsive cones in their eyes. This was once true for us as well, until 35 to 30 million years ago, when our own primate ancestors, still committed to life in the treetops, evolved a third color cone that allowed us to see with greater clarity, precision, and speed. (Color is perceived in the human brain just a bit more quickly than line or shape.) We're not as alert to dangers in the distance (if you live life in the trees, you usually can't see beyond the forest), but with three color-responsive cones and 20/20 vision, we can detect distant objects rather well. Horses, with 20/30 vision, can also detect objects in the distance, but not with our clarity.

Despite their differences, though, horse and human eyes share an important quality: they communicate emotions. For example, a horse eye that's showing the sclera, the white of the eye, or a human eye that's showing more of the sclera than usual, is communicating fear. We, too, read eyes in order to understand emotions. Charles Darwin considered this an innate skill that crossed cultural and even species boundaries.

Decades ago, the research psychologist Paul Ekman decided to create an experiment that would show whether Darwin was correct or not. Ekman traveled the globe taking photographs of people from remote cultures that had had little, if any, contact with the world at large—something that would be rare indeed in the twenty-first century—and found they could easily read the emotions of people they'd never encountered.

Recent research has shown that this ability may be based in certain areas of the brain, and may be inherited. Our brains have at least five different areas called "face patches," as explained by the Nobel

Prize–winning neuroscientist Eric Kandel in *The Age of Insight*. These groups of brain cells are dedicated to reading and responding to the facial features and expressions of others, confirming Ekman's finding that facial expressions are a sort of universal, inherited communication system. Interestingly, at least one of these patches of cells connects directly to the amygdala, the structure that mediates our emotions and social behavior.

Now, research by ethologists and animal behaviorists is confirming that this inherited ability may not be limited to *Homo sapiens*. Certainly, some dogs are able to do this, and as any horse owner will tell you, horses, too, can read eye and facial expressions.

The Canadian researchers Ian Whishaw and Emilyne Jankunis have found that horses have a universal expression they make when tasting something sour, and that this expression is quite similar to the expression we make when we take a big gulp of pure lemon juice. The two researchers gave forty-four horses sugar water and quinine. The horses who received the sugar water bobbed their heads, perked their ears, and slightly licked their lips. The horses who received the bitter infusion drew their lips back in distaste and stuck their tongues out—much as a child would do when asked to take an unpleasant medicine.

As Ekman showed, we are preprogrammed to read these expressions. Charles Darwin first proposed that we can read the intentions of other animals in this way because that ability helps us survive. Sometimes we do this not because of our experience, but in spite of it. Years ago I was canoeing in Zimbabwe through an area filled with hippopotamuses. My only experience with these animals was in children's picture books, when the laughing hippo opens his mouth charmingly in the middle of an African lake. In my own canoe, however, when a hippo rose up out of the water and opened his mouth at me, I understood immediately that he meant that as a threat. He was neither laughing nor charming, and I was not amused. I knew instantaneously that I was in danger. No words had to pass between me and my paddling partner. We just both immediately paddled backward.

In *The Expression of the Emotions in Man and Animals*, Darwin discussed the similarity of emotional expression in the eyes of many different mammals, and theorized that a set of universal emotions coupled to these expressions had evolved over tens of millions of years, just the way skeletons had evolved. Darwin was an excellent rider—his father often chided him for wasting time galloping his horse all over the countryside when he should have been figuring out how he was going to earn a living—and he understood the emotional expressions of horses quite well. As an older man, he once recounted a terribly dangerous ride when his horse encountered an unexpected tarpaulin lying in a field and communicated his terror directly to Darwin: "His eyes and ears were directed intently forward," Darwin wrote long after the event, "and I could feel through the saddle the palpitations of his heart." Years after the incident, his writing about it carried a vividness that showed that Darwin himself had never gotten over his own fright at nearly being dumped on the ground.

The clinical psychologist Sandra Wise believes that the dance of communication between horses and humans depends very much on the ability of both horses and humans to exchange information by way of eye contact. "In the eye of a horse, you can see a yin and yang," she told me. "You can see their genetically based fear, since they're prey animals, but you can also see their natural curiosity." Wise runs a program that teaches the finer points of human-human interaction by having students interact with her free-roaming Florida Cracker horses, horses that hark back to the earliest days of Spanish colonization. In her program, Eye of a Horse, students learn to watch the horse's eyes carefully in order to understand how they themselves are affecting the animal. Her clients include people with autism, people who have experienced personal traumas such as childhood abandonment, and masters- and doctoral-level students in clinical psychology. She believes that people who learn how to bond with a horse by making productive eye contact will have taken an important step toward being able to bond with other people.

Horses' eyes are huge, larger than even the eyes of elephants. The size of the eye may be related to the horse's defensive strategy of running from predators. The biologist Christopher Kirk has compared the top running speeds of a number of animals with the size of those animals' eyes. He theorizes that the horse's eye is so large because the horse can run forty miles per hour over an open plain in short spurts. Because of this need to see in the distance while running, Kirk suggests, horses have eyes that respond quickly to the world around them.

"How did humans and horses evolve their special relationship? I think it has something to do with their eyesight," Kirk told me. Of all non-primate mammals, he said, horses have visual abilities that are most similar to our own. "The eye is just a bit smaller than a racquetball," he said, "and they have tremendous acuity." By "tremendous" he meant in comparison to most other mammals, and closer to the exceptional acuity of primate vision.

For us, making eye contact is essential to our mental health. We do this persistently even with other animal species. Trained to herd sheep competitively, my border collie was taught *not* to make eye contact with his shepherd—who ended up being me. I found this very disconcerting, and felt rewarded when he finally made eye contact with me. He's not as good at herding sheep as he was when I got him, but he's a much better life companion for me. As a species, we crave eye contact. Psychologists have found that infants need to make eye contact with their caregivers in order to thrive. As adults we retain this fascination. We are, for example, suckers for the wide eyes of infants. We can't help ourselves. This is our biology. Looking into the eyes of other living things is almost addictive.

Who better to make eye contact with than the horse, the land-based mammal with our planet's largest eyes? This usually unconscious need has a great deal to do with our natural attraction to horses. Artists know this. Since the days of the Vogelherd horse, they have created horses with large, limpid eyes. In *The Arab Tent*, the British Victorian artist Edwin Landseer depicted an Arab mare and foal lying together on an elaborate carpet just inside a desert tent. When we see the soulful, huge black eyes of the gentle white mare, our hearts sigh. In Raphael's *Saint George and the Dragon*, the wide-eyed horse looks gratefully at his rider after Saint George has slain the predator. In 1769, the British artist Sawrey Gilpin painted *Gulliver Taking His Final Leave of the Land of the Houyhnhnms*, in which the eyes of horses express disdain for the human, whom they are about to throw off the island because of his lack of intelligence. In George Stubbs's *Mares and Foals*, a mare glares a warning at another horse who has come too close to her foal.

In the early twentieth century, Franz Marc's peaceful horses often have their eyes half-closed. The effect is transcendent. Walking through the Viennese early eighteenth-century stables of Prince Eugene of Savoy, a famous warrior skilled at using cavalry in his battles with the Turks, I saw that his favorite horses drank from marble water troughs, and I looked up to see benevolent horses' heads, also carved of marble, looking down protectively at me.

I wondered what made these horses seem so kind. Then I realized that the horses had been given impossibly enormous eyes, eyes that had been moved so far forward to the fronts of their skulls that their faces looked quite human. They reminded me of the cherubim in the Sistine Chapel. These horses were clearly gifts from the gods. Who wouldn't love such animals?

But the eyes of the dawn horses were not at all "kind." In fact, they were rather rodent-like, located halfway down the skull toward the muzzle. To think more about the evolution of the horse's eyes, I returned to the American Museum of Natural History, where the paleontologist Matthew Mihlbachler once again patiently pulled bones and more bones out of storage cabinets. The exhibits in the public part of the museum had changed several times since I had last visited Mihlbachler, but these back rooms looked pretty much the same. I sniffed the air. There, once again, was that aroma: dust and eternity.

This time Mihlbachler led me first to the department of modern mammals. He brought out skulls from modern *Equus* species. These include the horse, the plains zebra, the mountain zebra, the onager (an endangered species who lives in some Asian and Arabian deserts), the donkey, and the Przewalski's horse—all of whom descended from the original *Equus simplicidens* of 4 million years ago.

We looked for the location of the eye socket in each of these skulls. At the La Brea tar pits, Eric Scott had shown me a skull that came from a horse pulled out of the tar and compared it to a modern Thoroughbred. The racehorse had very forward-facing eyes that would have provided fairly good binocular vision, but would have handicapped the horse had he lived in the wild, since he would have had difficulty seeing predators sneaking up behind him. The ancient La Brea skull, on the other hand, had eye sockets located well to the sides of the horse's head. He would not have had very good binocular vision, but would have been able to keep one eye out, looking for danger on the far horizon, while he kept grazing. We were curious about whether we would see such differences from species to species, but they didn't stand out.

Then we wondered about how the placement of the horses' eyes had changed over time. We walked back 56 million years by heading over to the paleontological section of the museum. This research facility is monumental. The trek from the Holocene all the way back to the Eocene turned out to be quite a long journey. Traversing halls as long as New York City blocks, we descended cavernous stairwells that smelled of 150 years of scientific research, passed through heavy fire doors, and rode an elevator shaft, until finally we were back where I had first met Mihlbachler, on the sepulchral floor that held the horse fossils.

Once again, Mihlbachler pulled out skulls—this time, of several long-extinct horse species. He laid the skulls down on a long table, the kind you find in a school cafeteria, and formed an evolutionary tree of horse heads.

A lineup of modern Equus *skulls at the American Museum of Natural History* (Greg Auger)

At the base of the tree were the dawn horses; rising up from that foundation, he created branches to show how many different options horses had chosen over the course of their existence on Earth.

Suddenly, horse evolution was very real. By now, I had seen the horse's evolutionary tree time and again in scientific papers, but there on the table was the real thing: the change in horse skull shapes over tens of millions of years. While the jawbones of horses, containing mineralized teeth, are common all over the Northern Hemisphere, complete skulls are delicate and rare. Jawbones of animals often survive, Mihlbachler told me, because the teeth in the jawbones are tough and act as a kind of glue that keeps the bone from shattering. The plate bones of the skull, on the other hand, are easily broken and are thus difficult to find.

As I looked at Mihlbachler's tree of skulls, I thought once again of sickly, anxious Charles Darwin who was so often afraid to publish his books. Chances are good that if Darwin had been able to see what we were looking at that day, his stomach would have bothered him less and he would have saved himself a lot of trips to expensive health spas.

Mihlbachler pointed to a dawn horse skull.

"That animal was probably nocturnal," he said. "He definitely had large eyes, compared to his size."

He also pointed out strange indentations in a skull laid out on the table. These indentations sat just below the eye sockets and are called "preorbital fossae." They are not present in modern *Equus*.

"What were they for?" I asked.

"We have no idea," he answered. "It's kind of a big mystery."

Some scientists have suggested that the indentations contained scent glands; other researchers, that they were locations for certain muscles.

The eyes of the dawn horses were located well to the sides of the skull and nearer to the muzzle than the eyes of *Equus*. The dawn horses' eyes sat just above their teeth. As we looked at the change in horse skulls over time, we could track the spread of grasslands by the change in eye socket location.

"As the horses' teeth got larger," Mihlbachler explained, "the eyes had to move closer to the ears to make room." Relocating the eye sockets allowed for heftier molars that could consume more silica-laden grass.

So, I realized, the evolution of the horse's eye is connected to the evolution of the horse's teeth, and the evolution of the teeth is connected to the evolution of grass, and the evolution of grass is connected to changes in global temperatures, and the changes in temperature are connected to tectonic movements and changing ocean currents and the tendency of Antarctica to go it alone by reigning over the South Pole.

What we see by looking into the eyes of the horse is that we are all members of one constantly seething energy system.

So we understand now the truth that Darwin struggled with: our world is about change, and the corollary that not all change is equally successful. Some ideas—the colossal cecum of the horse, for example—turn out to have long-term staying power, while other ideas may work at one time, but turn out to have dire long-term consequences. Too much change too fast may create highly specialized animals who can't make ecological shifts when the world around them shifts in a major way, such as a sharp rise or fall in temperature. They may have simply become too highly specialized to survive. On the other hand, not changing at all has its own drawbacks. To illustrate that evolution is about synchrony with the natural world, Mihlbachler laid out on the tabletop a whole separate branch of the evolutionary tree of horses—one that also evolved large, fabulous animals every bit as exciting as modern horses, but one that ultimately disappeared from our world. Eventually, they didn't fit in anymore. There are many such branches in horse evolution, like the outstandingly successful *Hipparion*, who spread worldwide and then disappeared entirely.

Among the skulls on the tabletop was one from *Merychippus*, the horse that appeared somewhere around 17 million years ago and the horse from which *Equus* would eventually emerge. There was also a skull of *Megahippus*, a member of the dead-end branch we were thinking about. "If you saw this animal in a zoo," Mihlbachler said, pointing to *Merychippus*, "you'd say that it was some kind of weird horse. But if you saw this"—he pointed to *Megahippus*—"you wouldn't know what it was."

And he was right. I could see the future of the horse in that little *Merychippus* skull, but *Megahippus*, who lived only 10 million years ago, just before *Equus* appeared, was one bizarre guy.

He was large, like a modern horse, but he never evolved the rest of the horse package—long legs, or single-toed hoofs. He remained three-toed, which limited his opportunities in a world about to become very cold. His teeth never modernized, so he would have been stuck with browsing rather than grazing. Nor did the shape of his muzzle evolve to a wide form suitable for grass grazing. Instead, it stayed delicate and narrow, like the muzzle of a deer. Unlike the Yukon horse, *Megahippus* would not have been able to nuzzle through snow to pick at hidden bits of greenery. Nor would he have been able to see across wide grasslands. His eyes stayed halfway down his skull rather than moving closer to his ears.

Suited for forest life rather than open steppes, *Megahippus* died out. Some people call this "survival of the fittest," but that's just a popular perversion of what Charles Darwin meant: *Megahippus* fit the world in which he lived—but that world disappeared. It was really a matter of bad luck, more than anything else. Indeed, if *Megahippus*'s world hadn't contracted and the world of *Equus* hadn't expanded, we might be riding *Megahippus* today—although, given the odd shape of his backbone, we would need a different saddle.

Interestingly, while the eyes of the dawn horses showed only a bit of promise of what was to come, the eyes of the Polecat Bench primate already looked something like our own eyes. As Phil Gingerich told me, this primate's eyes had moved forward to the front of his face, and he probably already had fairly good binocular vision.

We needed all this binocular vision. It provides depth perception. Who wants to jump from treetop to treetop without knowing exactly how thick our destination branch is? Or precisely how far away it is? One mistake and we're down on the forest floor.

Our ability to see more colors than other placental mammals also aids depth perception. Seeing colors is a way to perceive specificity. When we evolved a third color cone, some time around 35 million years ago, a whole new world opened to us: we were suddenly able to distinguish "red." For those of us who can see red (some people are colorblind and cannot distinguish this color), it's hard to imagine *not* seeing it. But most mammals cannot.

This includes horses, who never evolved this third cone. When they look at a red object, they see color—but not the distinct red that we perceive. Most likely, researchers believe, our "red" is a yellowish-greenish hue to them. If we look at a red ball lying on green grass, the ball will stand out because of its color. If a horse looks at the same ball, the ball will not stand out. That's one reason why you may notice the ball at a distance, but your horse may only notice that ball when he is much nearer. And when he does notice it, it may startle him.

Our ability to see color is a key component of our ability to perceive the world around us, so that the fewer colors we are able to recognize, the less distinct our perceptions—so it's important to consider the world of color in which the horse lives. But just what is that world like? When I was a child, I was taught that dogs and horses see only in tones of black and white, but today we know that isn't true. It's likely that the range of color detection in horses and dogs is quite similar.

But how could scientists be so sure of this fact? Maybe our own "blue" color cone picks up on one shade of blue, while the blue cone of horses and dogs picks up on an entirely different one.

I asked the color vision researcher Joseph Carroll, who has studied this phenomenon in a number of animals. It turns out that we know a lot about color vision in humans, and, by proxy, in dogs and horses—thanks to researchers who wanted to understand the phenomenon of red-green color blindness, the inability of some people to detect the color red. One of the many high-tech tools that research has spawned is the electroretinograph, or ERG, which Carroll has used to understand color vision in animals. The device records the electrical activity in the eye rather the way an EKG records the electrical activity of the heart. "You flash different colors of lights and see how the eye responds.

If you do this over and over, you can deduce the color sensitivity," Carroll explained.

It turns out that the horse's sensitivity to color is quite like that of people who lack the third color cone and are red-green color-blind. To illustrate the range of colors horses see, Carroll took a photograph of two children wearing red clothing mounted on horseback. He took a second photograph of a horse walking in a paddock next to a barn with a green field in the background. Then he altered the colors in the photograph to simulate what a horse might see.

He also created a color wheel for horses. The green grass a horse sees is still "green," but not the rich, vital green most humans enjoy. Instead, the color is washed out. There are no reds. Blue is there, but lacks richness. From our perspective, horses see a washed-out world.

That missing vitality means that the horse has a limited ability to perceive fine detail. In Carroll's photographs comparing a scene as observed by the eyes of a human with the same scene as observed by the eyes of a horse, to our eye a brightly colored halter stands out on the head of the white horse. From the horse's point of view, that halter is much more dully colored and hence less obvious.

Gerald Jacobs, an expert in the evolution of color vision, believes that the ability of animals to detect color may have begun to evolve as long as 540 million years ago, when the first animals appeared in the ocean. "Color vision was probably inevitable," he told me. "Color vision is the way you resolve space."

If you have poor vision, it's easy to understand what Jacobs means. Take your glasses off and you will still perceive objects—not by perceiving sharp outlines but by perceiving color differences. You can navigate a room by following those differences. Now imagine that you are a horse. You'll have fewer color cues.

Horses, therefore, have more difficulty perceiving objects than we do. Imagine a red British public bus in the distance, behind a wall of green foliage. Although you can only detect bits and pieces of that red bus, in your mind you will "see" the red bus, because your brain will fill in the missing pieces for you. "The brain is a creativity machine that seeks out coherent patterns in an often confusing welter" of signals, as the neuroscientist Eric Kandel has pointed out.

For a horse, the red will not be obvious. He may not have enough information to perceive the bus. Imagine a few apples on a faraway, well-weathered picnic table. You see them and ride over for a snack. Your horse may not know they are apples until he gets close enough to detect the aroma. On the other hand, if he is used to seeing a tasty pile of apples on the picnic table, his mental creativity machine may put the pieces of the picture together and he'll head right over. This is why it's very important for horses to be allowed to experience new things in familiar areas. They need to examine novelty in order to assemble their own mental pictures.

This is particularly true for distant objects. Imagine a rider wearing a red jacket standing against a field of green grass. We can clearly see this person. The horse might not—until the person moves. At that point, he may well startle and bolt.

Sadly, although vision researchers can tell us about the colors a horse can detect, they can't tell us what the horse perceives, or how the horse puts the bits and pieces of visual information together into a mental picture of the world around him. But we do know that the horse is brilliant at assimilating novelty: once he's looked closely at something, he's likely to come to terms with it.

We know a great deal about this visual learning process in the human brain, and something about this process in the brains of dogs and even cats. But, as Gerald Jacobs warned, how the horse brain processes visual information remains a mystery, other than that the horse probably does not detect the color red.

That raised another question for me.

"Then what," I asked, "about the New York City carriage horse who stopped at the red light?"

Several years ago, a Central Park carriage horse named Oreo, frightened by an unexpected loud noise, had stampeded down one of the city's major thoroughfares, Ninth Avenue. Several people had tried to stop him, but Oreo was too frightened, having by then run into a silver BMW, torn the car's bumper off, and decimated his own carriage, still clattering behind him.

Then the horse came to an intersection. The light turned red. Oreo stopped, standing with vehicles waiting for the light to change.

"If horses are red-green color-blind, how did he know to stop?" I asked.

Jacobs explained that we can't know exactly why Oreo stopped at the red light, but it's possible the horse was aware of the position of the red light versus that of the green light. Horses may not distinguish the color red from the color green, but they can use all kinds of other visual cues in order to navigate.

He may also, Jacobs said, have seen a color difference in the two lights. That's because "green" lights in traffic signals are no longer true greens. Of course, the horse may simply have stopped because other vehicles had stopped at the light, but he may also have perceived a real color difference: "If the red and the green lights were perfectly red and perfectly green, and balanced in luminance [brightness], color-blind drivers would not be able to see the difference at all. What highway engineers did was add a lot of short-wavelength light to the green. It has a bluish cast, so it makes the discrimination one that could be detected by color-blind drivers who, like Oreo, only have two color cones."

And now, with a foundation for understanding how horses and humans see, we can return to Brian Timney and his research. After watching his horses watch the hot air balloons seven miles away, Timney decided to find out just how well horses could see when relying on acuity rather than color. In humans, acuity—our sharpness of vision—is central to our perception of the world. Our own acuity is tested when we read an eye chart: when we get to the line that we can't read, the optometrist tells us just how bad our vision is. Good human eyesight is 20/20. My own vision was 20/one zillion until I had eye surgery. Things that others saw in the distance simply did not exist for me. I was in a national park once when a number of people stood and watched a grizzly bear climb on a not-that-distant cliff. I saw nothing.

But how do you figure something like that out for a horse, since they can't tell us what they're seeing? Timney applied time-honored techniques for assessing the acuity of human infants, but added a few twists. First he built two hinged swinging doors into a wall-like structure. Behind each door were trays with treats. To get the treats, the horse had to learn to push the swinging door with his muzzle. As any barn manager knows, horses learn such things easily.

Next, Timney kicked things up a notch. The horses learned that there were no treats behind the door painted solid gray. There *was* a treat behind the door painted in very wide black and white stripes. The horses also learned this easily, so he knew the horses were "reading" the stripes.

Then he began progressively narrowing the width of the stripes. This gradually increased the difficulty of detecting which door was striped and which was not. In a sense, the horses were now reading a version of an optometrist's eye chart.

To ensure that the test measured acuity at a distance, Timney had the horses stand back a bit more than six feet. A two-meter-long wall separated the path leading to one door from the path leading to the other door, so that the horses had to make their choices from a distance.

The horses were pretty good at the game. At first, when the stripes were quite wide, their success level was nearly 100 percent. But as the stripes narrowed in width, getting closer and closer to solid gray, the horses had more and more difficulty choosing. When they made correct choices only half the time, Timney knew they were just guessing. The very narrow stripes were probably appearing to the horse to be no different from the solid gray.

This very cool experiment tells us a lot. First, it reveals that horses make good research subjects, as they were willing to stand and learn in order to get a paycheck. It also tells us that although horses have vision that's less than 20/20, they do see with more clarity than most mammals other than primates. Timney has conducted similar experiments with cats, monkeys, camels, and even bumblebees. He's found that the horse's long-distance acuity is superior to that of most other animals.

"Horses have a visual acuity that's about two-thirds as good as human acuity," he told me. "That's actually pretty good. Certainly, when they saw those hot air balloons they were picking up something that was pretty small. A cat could have had no hope of seeing something that distant." This tells us that not only does a horse have relatively good vision compared to a lot of other mammals, but that when a rider is on a horse's back and they're going for the jump, the horse *will* be able to see the jump relatively well.

Timney also wanted to know whether a horse can perceive depth. Was the horse's world simply flat and two-dimensional? Probably not, or horses would have difficulty jumping over fences. But what kinds of depth-related concepts do they have?

He used the Ponzo illusion to figure this out. This is a two-dimensional visual deception that fools most of us—so much so that even after it's explained, we still have trouble seeing what's really there.

Draw two lines of equal length on a piece of paper, stacked one on top of the other, so that they make a kind of column. You'll easily see those lines as equal in length.

But then put those two equal-length lines in context. If you place them on top of a drawing of a set of receding railroad tracks, you will perceive the top line as longer than the bottom line. Your eye will detect the data correctly, but when your brain assembles the data into a picture, your brain will misinform you.

This phenomenon is so strong and consistent that even when we know for a fact that the two lines are exactly the same length, we will continue to misperceive their relationship. The neuroscientist Kandel explains the phenomenon this way: "Vision is not simply a window onto the world, but truly a creation of the brain." In other words, the human brain assembles a concept of depth—even when it's not really there. When Renaissance artists finally realized this strange truth—that the brain automatically "sees" three-dimensional perspective even in flat, two-dimensional art—Western art completely changed.

Timney discovered that horses do the same thing. He showed his horses two separate sets of lines stacked on top of each other. One set of lines was of equal length, as in the experiment with people. The other set showed a line on top that was indeed longer than the lower line. The horses learned to go to the set of lines with the longer line on top. Then he presented his horses with the Ponzo illusion. He showed them two sets of two lines—but both sets contained lines of equal length. One set of lines was placed in a drawing with trees and other landscape items *not* drawn in perspective. The second set of lines was placed on top of the receding railroad tracks, as in the human test.

"Overwhelmingly, they chose to go to the one that seemed as though the line was longer at the top," he told me. "They were susceptible to the illusion."

This is stunning. A visual ability of which we humans are particularly proud, our ability to "read" depth on a piece of paper—turns out to be something that horses do, too. Moreover, the fact that horses and humans make the same error in perception provides yet another clue to our common evolutionary heritage. Our most recent shared ancestor may have had the same tendency. It made me wonder: What would horses make of Venetian Renaissance art, with its great advances in communicating depth? Would they be able to pick that painting out of a group of earlier paintings that don't communicate depth?

It's a silly question, asked lightly, but it is remarkable that horses can extrapolate information from a two-dimensional page clueing them in to facts about the three-dimensional world around them. If you think about what the horse is doing, it shows a pretty sophisticated mental capacity. Is their sense of perspective similar to our own?

I wouldn't have thought so. And yet, Timney's research shows that there may be more similarities than we expect. After all, life evolved in a three-dimensional world. It makes sense that brains evolved to somehow deal with that fact.

Excellent binocular vision is a terrific aid in depth perception, but there are other ways in which animals can perceive depth and distance. We humans have at least one other tool in our toolbox. Open one of your eyes and move your head: you'll easily perceive which objects are near to you and which are far away. This is because of a phenomenon called "motion parallax." Looking out the window of a speeding car with one eye closed, you will still know which objects are close to you and which objects are farther away.

Horses do this, too. They can do this when galloping or walking, but they can also do this just by moving their heads. This is one reason why it's important to let a horse move his head when you're riding. Holding a horse on a tight rein inhibits the horse's ability to use motion parallax to perceive depth. For example, with one eye, a horse may have trouble assembling a mental picture of the world around him when the light is dappled, as it is under a canopy of trees. We can use our binocular vision to do this, but horses have limited binocular vision and thus more trouble with this task. This is yet another reason why the horse responds to any movement, any sparkling difference between light and shadow, and not just to specific objects.

We humans excel at color detection, compared to other mammals, but horses are much better than we are at detecting even the smallest movements in low light, as at dusk. This is because horses, compared to us, have a greater proportion of super-light-sensitive rods to every color cone. There seem to be wiring differences as well. Some rods in horse eyes send messages to the central nervous system at much greater speeds than do the rods in our own eyes.

On the other hand, our eyes adapt more quickly to changes in light levels. If you turn out the lights in a room at night, you'll be able to see in a matter of seconds. The horse requires about a half hour to make the equivalent change. Because our primate ancestors lived in thickly forested areas, we needed this dark-to-light-and-back-again flexibility. Evolved to live on the open plain, *Equus* only required rods that changed during sunset or sunrise, or about the evolutionarily appropriate thirty minutes.

We often put horses in situations where they cannot see very well. When we lead a horse from bright sunlight into a dark horse trailer, for example, our own vision improves immediately. But the horse's vision will not be good for about thirty minutes. To him it may seem as though we are leading him into a dark, dangerous cave.

Sometimes when a horse appears to be behaving badly, the problem is not the horse per se, but his difficulty in seeing. Outside of Edinburgh, Scotland, I once saw a trainer bring a "problem" horse out of the sunlight into a poorly lit arena. All alone, with only a stranger holding his lead line, the horse was standing very much on alert. Every muscle in his body was tense. His head was held very high and his nose was up in the air. He appeared to be trying to see something. His ears were pointed directly forward.

Following the direction of his eyes and ears, it was easy to see what frightened him. The far end of the arena was darker than where he was standing. At that end was a solid wall that stood out visually since it was painted a bright, luminous white. This solid wall rose to about shoulder height. Behind that white wall people in black watch caps scurried back and forth. Their bodies were hidden. Only their black-capped heads were visible. To us, because our acuity is better and because we've had experience with watch caps, the wall looked like a wall and the people looked like people. Our brains created an appropriate picture.

The horse, a newcomer to the riding arena, was unable to put the mental picture together. He probably perceived something that his evolutionary heritage predisposed him to notice. Perhaps the picture the horse assembled from the data gathered by his eyes was not a picture of a white wall, but a picture of a faraway *cliff.* And upon that white cliff were black-pelted predators—wolves?—scurrying back and forth, biding their time until ready to pounce. No wonder the horse was nervous. The poor thing kept snorting, dancing back and forth, and raising his head, trying to get a better look.

The horse behaved like this *only* when alone in the arena, a trainer said. When other horses were with him, this horse was quite calm. As we know, horses travel in bands, and like other mammals who live in tightly organized groups—prairie dogs, for example—they often rely as much on the vision of their band members as on their own vision. That's why horses in the open rarely all lie down at the same time. One has to remain on guard.

Possibly if the skittish horse had been with a longtime, trusted human companion, he might have been calmer, but sometimes horses may rely too much on human cues. The Australian neuroscientist Alison Harman, an experienced dressage rider, once saw two dressage horses, with noses tucked into their chests, canter toward each other and collide. Harman conducted research that revealed that horses with heads tucked in to their chests are traveling blind. The horses who collided were depending on their riders' cues, and their riders were not paying attention.

Harman discovered that horses have a blind spot directly in front of their faces. Because of their evolutionary heritage, horses see well down the length of their noses, so that they can see which plants they are eating in a field, and they can see almost (but not quite) directly behind them, so they can always keep an eye out for predators slowly creeping through tall grasses. But they cannot see in front of their faces the way we can. If you're brushing in between a horse's eyes, for example, he can't see the brush and may only catch brief glimpses of your moving hand. He has to learn to trust you and your brush.

Harman also showed that the world the horse sees is wide and flat and narrow. Most of the horse's color cones and many of the rods are located in a thin visual strip that runs all along the retina. There are only a few rods and cones located outside of this visual strip, while the strip itself is so thickly populated with rods and cones that the horse pays attention to the world in a strip-like fashion.

By contrast, our own vision is circular. When we take a jump, we see that jump in the center of our visual circle. When our horse takes the same jump, it's just one of many objects he perceives. He is paying attention to movement across the whole of that visual strip and responding to anything that, in his experience, shouldn't be there. He's particularly sensitive to movements picked up on the edges of that visual strip. We don't see *well* out of the corners of our eyes, and neither do horses—but they are wired to respond immediately to any movement picked up that way, regardless of what it might be. This is one reason why carriage horses wear blinkers—the flashing of the turning spokes, picked up in the corners of the horse's eye, signals something dangerous following him, but he cannot perceive what that danger might be.

A panoramic view of nearly 360 degrees means the horse is taking in an awful lot of data. How does he handle what might seem to us like visual overload? Most likely, he does not treat all the data the same, but instead picks out various pieces of information—particularly novelty—to which he responds. Given enough time, horses can learn to tune out unimportant visual information.

We do the same thing. When we drive a car, for example, we don't see much of the world we're passing through, but we do see what's important, like a car ahead of us veering erratically. We even do this aurally in a busy restaurant. When we're trying to listen to a conversation amid the clatter of knives and forks and the chatter at other tables, we ignore most sounds and only perceive the desired words.

Similarly, the horse may well filter out most of his visual world and "see" only what matters most to him. Just as we don't pay attention to much of what's out there when we drive but do notice a car careening toward us because that's important to us, the horse is predisposed to notice unexpected movements that, because of his evolutionary heritage, may be important to him—the rustling of grass behind him, for example, or a small movement up on a distant ridgetop.

When I was a child learning to ride, no one ever talked to me about how horses perceive the world. It was Whisper and Gray who taught me the importance of taking into account our visual differences. One cool October morning, when crisp Canadian air had arrived overnight to turn the maple, oak, and birch trees along my dirt road into a thousand shades of glory, it seemed a fine time for a ride. I tacked up Whisper and put a Western saddle on Gray. A friend who didn't ride wanted to come along. My old half Percheron, always staunchly opposed to any kind of physical exertion, seemed an excellent mount for her.

About five minutes down the road, we saw something brand-new in my tiny town: a high-stepping pony pulling a smart two-wheeled buggy. The spokes of the buggy's turning wheels sparkled brightly in the sunlight speckling its way through the dark, cathedral-like forest canopy.

I took note. I knew nothing about horse vision, but I did know that horses do not like novelty. But both Whisper and Gray seemed nonchalant. The buggy came closer. The pony trotted toward us. The horses paid no attention. Copacetic.

Then, when the cart was about forty feet away, Gray bolted. Like a Thoroughbred busting out of the starting gate, he tore along in high gear with his neck stretched out. It was useless to tell my friend to pull back on the reins. Even if she'd dared to let go of the saddle horn, her efforts would have been futile. Gray was a horse on a mission, and his mission was to get away from the monster that had, without warning, loomed up in his field of vision.

Whisper, of course, saved the day. He caught up with Gray, got in front of him, slowed to a jog and then to a walk. Gray calmed down. My friend did not. As far as I know, she's never ridden again.

It was a good life lesson for me. Horses do not see what I see. But I always wondered: What did Gray perceive that fall afternoon? When did he notice the cart and pony, and what in the world did he think he saw? What could possibly have been so terrifying?

Science has now provided me with a few answers to the questions I had back when Gray bolted. I know that when Gray and I looked up into the fall sky, we both probably saw the color blue, although the blue that Gray saw may have been less stunning than the rich, deep blue I enjoyed. When we both looked at the autumn leaves, the beautiful hues that I saw were probably only barely perceived by the horse.

And even though the sun was shining, the tall trees on each side of the road made much of the road ahead quite dark, so that the horses might have had trouble seeing into the distance. When the buggy appeared, I knew what it was: my brain put together a picture of a smartly outfitted pony pulling a buggy. Gray may have perceived something quite different. As the buggy got closer, I saw more and more details, like the turning spokes and the pony's high head carriage. Gray probably did not see many of those details at first. And when the pony finally became clearly visible to him, his eyes may well have been alerted by the flashes of sunlight playing on the turning wheel spokes. These flashes may well have been even more difficult for him to cope with because of the speckled light coming through the treetop foliage. Did he finally put together a mental image of a pony being chased by a predator with bright, flashing eyes?

We can think about this, but we may never know for sure. Although science has in recent years found out a lot about how the eyes of the horse function, we don't know much about how that information is *perceived* in the horse's brain. How do horses assemble mental images of the world in which they live? We know something about how cats do this, because cats are small and make good research subjects. And we know something about how dogs do this, because we care about dogs and are willing to invest research money in studying their behavior. But horses are inconveniently large research subjects, and, although we care about them, we have been remiss about improving our scientific understanding of how their minds work.

In fact, although we've studied thinking, or cognition, extensively in a variety of animals ranging from whales and dolphins to cats and dogs, we are only beginning to scientifically explore how a horse thinks and how this might be connected to why he has been so willing to partner with humans.

9

THE DANCE OF COMMUNICATION

Under a transport of Joy or of vivid Pleasure, there is a strong tendency to various purposeless movements, and to the utterance of various sounds. We see this in our young children, in their loud laughter, clapping of hands, and jumping for joy; in the bounding and barking of a dog when going out to walk with his master; and in the frisking of a horse when turned out into an open field. Joy quickens the circulation.

—CHARLES DARWIN, *The Expression of the Emotions in Man and Animals*

In a paddock about an hour inland from the muddle of Los Angeles, an old ex-racehorse and his owner were dancing up a storm. Their elegant ballet was as evocative as any Pleistocene painting, as vital as the Vogelherd horse.

Karen Murdock raised her arms in an exquisite expression of exultation. Lukas reared gracefully, holding up his side of the interchange. In response, Karen waggled her finger. Lukas backed up. She signaled him to lope, she signaled him to jog, she signaled him to come hither and then again to go away.

The exchange went both ways. Sometimes Lukas, a tall chestnut, initiated an action and Karen responded. Lukas turned out to be an expert at getting Karen to smile, and by now he knew that her smile was likely to be followed by some other kind of engaging behavior. They were an old couple, and they knew each other very well.

Karen and Lukas (Joan Malloch, courtesy of Karen Murdock)

In the midst of their exchange, Karen turned away from the horse and spoke to me about the two-way nature of their partnership. "These are not 'tricks.' Our whole lives go into this interaction. It's an entire process." She meant that she does not command and wait for Lukas to obey, but that the pair share equally in their interactions.

She meant "whole lives" almost literally: every day Karen leaves the suburban Southern California home she shares with her husband and heads for the barn, where she spends hours with Lukas, trying to learn as much as she can, in an informal fashion, about what makes horses do the things they do. It's a fascination she's had since childhood but can only indulge now that she's retired.

As Karen continued to talk to me, Lukas became impatient. He started using small, subtle movements to get Karen's attention. When those didn't work, he bowed his head between his knees. That was usually the clincher, but this time Karen ignored him. His restlessness accelerated. Ever more impatient with the small talk, he went on with the show, but nothing he did diverted Karen's attention.

Horses, like dogs, learn quickly that when they get a human to laugh, it's all over for the human. But this time, no dice. Lukas began to take more extreme steps. He tried the hokey pokey: one leg lifted forward, like a fancy dressage horse; then, the other. After that, the river of his imagination flowed forth and he offered all kinds of inventive body language. He stood with arched neck. He bent his head first in one direction, then in another. He tried walking away. No cigar.

Finally he walked over to Karen. Ever so gently he closed his front teeth over a tiny bit of her jacket and quietly but firmly led her off: enough is enough. We laughed, even though we knew we shouldn't have. Our laughter was partly due to the irony of a horse leading away his owner, but also because, like a toddler, Lukas had managed to end a conversation in which he had not been included.

Lukas's body language reminded me a great deal of the kinds of things I'd seen with Jason Ransom while watching the McCullough Peaks and Pryor Mountain horses in Wyoming, and of the patiently built partnership between Kris Kokal and his mustangs. Horses talk to each other a lot—"continuously," according to Ransom. Apparently, they never shut up. They talk to people, too, through body language, but it's rare to see humans respond in the way Karen does. This *was* a two-way street. The conversation went back and forth, like two old friends sharing cups of coffee.

To watch Karen and Lukas was enlightening. I've spent much of my life with horses, but never known that this deep level of exchange was possible. Karen finished what she'd been saying, that Lukas was not a "trick" horse, and that she was not a "trainer," but just someone who cared about Lukas: "I work with Lukas because I'm interested in bonding, because bonding has to come first."

Then she walked off with the horse, who had regained his position as the center of her attention. When they were finished waltzing and liberty time was over, we all walked back to Lukas's open-air stall. With Lukas inside and Karen outside, with just a rope between them, Karen opened a small folding table on which she put several large bright plastic numbers—6, 2, 5, 3—the kind you use to teach simple numbers to very young children.

When Karen called out a number, Lukas identified that number by touching the correct figure with his unshaven muzzle. When he answered correctly, which was most of the time, Karen gave him a carrot sliver. Karen and her husband spend well over an hour every evening slicing up the next day's supply. Karen likes to think up new activities that she and Lukas can do together, but she's not a scientist and is not doing formal research. Instead, she just wants to interact with her horse informally, to see what happens when force is not used.

Carrots may be the currency by which they live, but the bond between them goes much deeper. There is a great deal of debate in the horse world over whether it's "right" to reward a horse for doing something or whether the horse should perform the action because someone ordered him to, but it was apparent watching Karen and Lukas that their interaction was based on freedom rather than domination and that it had little to do with snacking. It was all about cooperation. The carrots were just the icing on the cake.

Folk wisdom says that a horse's attention span is quite short, but I watched Lukas spend several *hours* deeply engaged with Karen without getting restless. In fact, when Karen walked away for a few minutes, the horse looked steadfastly in her direction until she returned. Is this "friendship"? Is this what love is made of? Who knows?

There does seem to be a neurological basis for their connection. The neurobiologist Hans Hofmann has found that vertebrates as a group share a basic brain circuitry associated with bonding and reward processing. Certainly, what I saw was a very strong social bond, one that had blossomed over the years into a veritable marriage of personalities. Karen and Lukas enjoy the same activities, the same pastimes, the same games. You can almost imagine them at the same breakfast table reading the newspaper together. Karen has also written a book, *Playing with Lukas*, that describes their relationship. One of my favorite sentences: "Lukas and I are in the round pen together twenty feet apart—our eyes locked on each other. The world has faded away and time has stopped. Nothing exists but our gaze." I saw that, whether in the paddock or the barn, both watched each other closely.

One of their favorite activities is the counting game. Karen holds out a carrot, and Lukas is not allowed to take it until Karen counts to three. Sometimes she counts one, two, three. But sometimes she counts one, two, fifty-nine, twenty-six, ninety-eight . . . It may be many seconds before she says "three." You can see Lukas straining to control himself. Impulse control is not something we normally ascribe to high-strung Thoroughbreds, and yet, there it was, clear as day. He bends his neck and turns his head coyly, the better to keep watch. He wants that carrot. Every once in a while, his muzzle gets fairly close, but he catches himself. He won't reach for it until Karen finally says the right word.

Karen says Lukas holds the "world's record" for the most numbers identified by a horse in only one minute. She also says, laughing, that she doesn't know of any other horses who have yet competed for the title. She does suggest that Lukas's performance may be an indication that he really can "count." Maybe so. On the other hand, Lukas may simply be so well tuned to Karen's psyche that he's following her lead by watching body language so subtle that the rest of us don't see it. Karen herself may not be aware of these cues. Lukas is, though. His huge eyes don't miss a thing. Because horses live in small bands and depend for protection on other band horses, they are particularly sensitive to even the slightest shifts in posture—and in this case, his bandmate was Karen.

It didn't matter to me whether Lukas could count or not. I hadn't come to find out whether horses were cleverly disguised Einsteins. Nor did I care about whether Lukas followed orders—bend your head, stand up on your legs. I had come for the dance. Their waltz was the more interesting when you consider that Lukas was once a very troubled horse. My sense of him is that he would still be a troubled horse, were it not for the bond he had formed with Karen. I once asked Karen whether she enjoyed riding him. "Oh," she said, "he doesn't really care for that."

Lukas ran a few races as a two-year-old. He didn't win, but he did suffer two bowed tendons, fating him to spend the rest of his life in pain. Several people tried to train him as a riding horse, but either because of the pain or because of his temperament, those efforts failed.

What do you do with a horse who can't be safely ridden and who is permanently lame? He passed through the hands of several people and developed a reputation as dangerous. He carried that hot-horse stigma from boarding barn to boarding barn. No one knew what to do with him. He had a lot of pent-up energy, but because of his tendons he couldn't rid himself of that energy by doing what he'd been bred to do—run.

Eventually, he was abandoned in a small pasture, left alone to fend for himself. Someone saw him, emaciated and needing a vet. She brought him home and put an ad in a local journal offering him as a "project" horse—meaning a horse that someone would have to work hard to rehabilitate. Karen bought him and found she had an enigma on her hands. She tried dressage with him, but Lukas remained unpredictable and rebellious.

At that point, Karen began to look for alternatives. If she couldn't "train" Lukas in the conventional sense of the word, maybe she could apply techniques learned from her job. She had been a nurse in a psychiatric ward, a career that involved placing herself in situations that were "intense, volatile, unpredictable," she told me.

Just like Lukas, I thought.

She used the techniques she'd learned when working with the tougher hospital cases, the ones where saying no to a patient brings forth a powder keg explosion. I saw as she worked with Lukas that even now, long after her retirement from nursing, she is adept at using quiet, firm, consistent, gentle, and subtle shaping cues. This was the key to her success. I never heard the word "no," but as the pair played with their numbers on the foldable table, Karen worked hard at helping Lukas stay within his prescribed boundaries. At times the horse edged toward what might be called excessive enthusiasm, and when he started to get wound up, his right front foot often inched just a tad outside the stall. Each time, Karen picked up the hoof and placed it back inside. There were no reprimands. Not a negative word. No shoves. Just shaping and consistency.

Lukas is still a very excitable horse, as I saw when he first entered the paddock. Although he was clearly in pain from the bowed tendons, the bucks and the gallops and the midair twists were still there. But I could also see that Lukas is a naturally curious horse, like the horses Jason Ransom and I watched in Wyoming. When alone in the paddock, he began checking things out, even though he'd been there a thousand times before. He walked over to where a squirrel had scurried along a fence rail. Then he stood at attention scanning the horizon, wanting to know if anything out there had changed. He sniffed the water trough, put his nose into fence corners, and generally filled himself in on what had changed since his last visit, which had been only the day before. While I watched, he never really settled down and relaxed. He was always antsy, looking for whatever problems might be just around the bend.

But when Karen came into the paddock, he settled immediately. He walked over to her and invited her into their private world.

The dancing began.

According to scientists who study the evolution of emotions and of behavior, the biological basis for this kind of deep understanding between two seemingly very different species—this nurture that's rooted in nature—comes from our shared natural history. "The difference in mind between man and the higher animals, great as it is, certainly is one of degree and not of kind," Darwin wrote in *The Descent of Man.* For Victorians, this was downright risqué. To equate the mind of humans in any way with the mind of other animals was pushing the envelope.

Now we know that Darwin was half right and half wrong. In categorizing animals as "higher" and "lower," he was buying into the Victorian belief that humanity was the pinnacle of evolution. Today we think about this differently. For example, we know that *Equus* is not "higher" than the long-gone *Megahippus*, but just better suited to the ecosystem in which he evolved. Today, rather than thinking about a ladder of life, we think about a web of life or even a jigsaw puzzle of life—to get the whole picture, we need to have all the pieces. We humans are indeed "special"—some of us can create great art or complex symphonies or even invent calculus—but we are also part of a system of interdependencies that includes factors like an absolute reliance on simple bacteria, which allow us to digest our food. We depend on other complex animals, who perform important functions in the world and who are, like us, dependent on intricate webs of energy systems—

systems involving tectonic forces, ocean currents, the rising and falling of global temperatures, and many other basic elements.

Without all that complexity—even the catastrophes—we would not be what we are today. When he wrote his great treatises, Darwin was only beginning to grasp these processes—a solid understanding of plate tectonics wouldn't come for another hundred years—so it's not surprising that although he realized that life changed over time, he continued to see life as hierarchical. When Darwin was alive, an inordinate number of those puzzle pieces, like plate tectonics, were missing. He was trying to figure out the whole picture with only a few bits of information. Nevertheless, genius that he was, he recognized the essential idea that life ebbs and flows with the passing of time and devoted his whole self to making that case. Imagine what he would have been able to understand had he only known that continents move.

Fortunately for us, over the past 150 years or so, scientists have managed to discover a few more clues to the full picture. There are certainly more out there, and when they are uncovered, our grasp of evolution will become even more fascinating. Darwin saw only the tip of the iceberg, and now we see a few feet below the surface. There's so much more to understand. I once read an essay written by a scientist at the end of the nineteenth century about the recent discovery of the electron. It's time, he concluded, for science to close up shop. We now know all there is to know. Then, only a few years later, in 1905, Einstein published $E=mc^2$, providing us with a whole new depth of understanding about the nature of energy.

Darwin knew quite well that his theory was just the start of a fabulous journey, and he might have been delighted to learn that evolution is not about who is "higher" than whom, but about how our *roots* connect us. Nature forms the foundation for nurture, like the pedestal on which a great piece of sculpture is placed.

Today we know that nature provides the fundamental commonalities that allow us to understand other living things. We understand the expression of fear in a horse's eyes because we share an evolutionary history with the horse. This doesn't mean that we are nothing *but* our biology, which is unfortunately how explanations are sometimes framed by some scientists and scholars who are not quite careful enough with their words. But it does mean that these biologically based commonalities vastly improve our lives, because we can understand the horses and other animals who share our world with us. Without these companions who have traveled with us through time, we're just floating in space, disconnected.

"Human beings are species-lonely," the author Thomas McGuane wrote in *Some Horses.* When I read his sentence, I felt grateful. Someone had finally voiced what those of us with horses have always known: there's something innate in us, as in horses, that yearns to bond. We must have horses and dogs and cats and other animals in our lives in order for our psyches to function as they should, just as we have to have bacteria in our guts in order to digest food.

Today, science is explaining something about why this is true. For example, cortisol is a hormone common to most animals. Even fish have cortisol, which tells us that it appeared early in vertebrate evolution. High levels of cortisol have long been known to correlate with high levels of stress and with stress-induced illnesses. Now research shows that people who have pets tend to have lower cortisol levels—indicating lower stress levels—than people without pets. People who spend their lives around animals often live longer, happier, healthier lives.

Pleistocene artists must have known that, too, on some level. I realized that when I visited their dwellings and saw their work. In the Spanish caves where people gathered in groups, horses and other animals on the cave walls kept the people company. The artists expressed the essence of these horses—"the *rasa* of a horse," to quote the neuroscientist Vilayanur Ramachandran—through simple lines showing the arch of the neck or the distinctive, graceful topline of the horse's back. It was as though these artists wanted the animals to always be with them, to keep them company.

This is innate. Years ago, in a tent by a river in Zimbabwe, I woke up at first light because the noise from the animals—chuffling hippopotamuses, screaming baboons, singing birds—was louder than early-morning garbage trucks on a New York City street. The cacophony was overwhelming. But unlike the noise made by city trucks, these noises were comforting, like the extremely loud, overwhelming *Freude* chorus of Beethoven's Ninth Symphony. The animals around my tent that morning were performing the ultimate "Ode to Joy," and it was one that, although new to my ears, I recognized.

Had Darwin managed to see the fruits of his thinking in the twenty-first century, he would have been, I believe, thrilled to learn that the mind of the human is not "superior" to the mind of the horse, but is, rather, complementary.

What was going on in Lukas's mind while he was dancing with Karen? We have to be careful here, when asking this question in a scientific manner. In the field of animal cognition, the story of Clever Hans, a horse from the early twentieth century and one of my all-time favorite horses, has become a cautionary tale about not reading too much into the mind of the horse. Clever Hans was a great horse. Like Lukas and Whisper, he managed to lead an interesting life. He and his owner were obsessed with each other. His owner taught the horse to do "math"

Clever Hans being tested

and frequently exhibited his skills in public. "How much is six plus two?" the owner would ask. Then Hans would tap eight times with his front hoof. He always gave the right answer. He turned out to be equally talented at multiplication and division.

Hans was famous. His face was plastered all over the front pages of newspapers worldwide. But, of course, there were people who doubted that Hans really was the genius his owner claimed he was. In response, Hans's owner agreed to a test. Instead of the owner taking Hans through his mathematical paces, others—strangers to Hans—gave him problems to solve. And still Hans answered correctly.

Then Hans's questioners hid behind a screen, so that they could test the horse without being seen. Suddenly, Hans could no longer respond. It turned out that Hans, like Lucas, was reading body language: when Hans's owner and other people asked questions, they moved their heads forward slightly when the horse got to the correct answer. Their body language was subtle and unintentional, yet universal to everyone who tested Hans. Remarkably, Hans was interpreting those tiny head movements. He understood that body language even when the testers were strangers to him. Hans, it seems, knew more about us than we knew about ourselves, at least in this regard.

Sadly, once the truth was known, Hans and his owner fell from grace. Because Hans couldn't really do math, people dismissed his abilities, disregarding the horse's obvious intelligence—his ability to understand what humans wanted by studying their behavior.

If Hans's true talent was overlooked, that was not unusual. Just as Charles Darwin saw evolution in terms of hierarchy, until very recently many of us have thought about intelligence in terms of what *humans* believe to be intelligent—like the ability to do math. Indeed, animals were not thought by some scientists to possess intelligence at all, but merely to behave according to the simple rules of reward and punishment—simple positive and negative reinforcement. The minds of all animals, including humans, were thought at birth to be blank slates upon which experience wrote. To attribute the ability to think to an animal was said to be just anthropomorphism.

Today we know that's not true, and that this black-and-white view was a handicap. Once we let go of that belief, we began a renaissance in our understanding of animal minds. For example, in the opening chapter of his groundbreaking book titled, appropriately, *Animal Minds*, Donald R. Griffin wrote that "conscious thinking may well be a *core function* of central nervous systems"—a statement that brought him a lot of heat when he first published his book in 1992. Not that Don cared: He was a natural-born revolutionary, a true outlier. His earlier book, *The Question of Animal Awareness*, published in 1976, suggesting that animals were conscious, was called "subversive" by one opponent, which delighted him. He was never one for conformity.

If he were alive today, he would have enjoyed seeing that his ideas led to an awakening. Researchers curious about animal thinking began constructing scientific experiments that would help us, slowly, penetrate the complexities of how other animals think. Unfortunately, he never addressed the mind of the horse, most likely because there was little extant research on that subject during his lifetime.

I attribute that neglect to two factors. First, during Don's lifetime, horses, for perhaps the first time in the history of the horse-human partnership, played a very small role in human culture. Cars had replaced horsepower, and horses had come to be seen in many places as luxuries. The invention of tanks and other military equipment meant that horses were no longer used in war. Farmers gave up horse-powered plows for tractors. Because horses were less common, few researchers were interested in them. But the more important factor was the story of Clever Hans. Anyone studying the mind of the horse was laying himself open to the dreaded charge of anthropomorphism.

That's changed over the last several decades. As the influence of behaviorism has waned and our understanding of evolution has improved, interest in studying the minds of animals has accelerated. Today a small group, based mostly in Europe, has begun to include horses in the array of animals whose ability to think is worthy of research. What these studies reveal in terms of the horse's intelligence is fascinating.

Of course, some people who are intimately involved with horses take for granted that they are intelligent. But there's a difference between anecdote and science. When I was young, I recognized that there was something special about Whisper. I was quite proud of his achievements. But those stories were just stories—not methodical investigations into his abilities.

Science often begins with an anecdote: Brian Timney saw his horses staring at hot air balloons in the distance. But then, based on this anecdotal experience, he devised a set of research experiments that revealed a great deal about how horses see. This is now also happening in the field of equine cognition. Some of the findings of these scientists are enlightening.

The Oklahoma research psychologist Sherril Stone decided to find out whether horses could learn about the three-dimensional world in which they lived by looking at a two-dimensional board. Brian Timney had already shown that horses, when looking at lines on a board, are susceptible to the Ponzo illusion—the misreading of line lengths placed on a flat board. Stone took this research one step further.

First she taught several horses to approach a particular shape, like a star or a triangle, shown in two dimensions on a flat board. She put two of these signs up on a wall. If the horse approached the correct sign on the wall, he received a food reward. If the horse approached the other sign, he did not.

Next, Stone made three-dimensional models of those two-dimensional shapes. She found that when the horses were released into a paddock, they went and stood by the three-dimensional shape that looked like what they'd seen two-dimensionally on the board. Most of the horses did pretty well with the task: they stood consistently by the three-dimensional object that resembled the two-dimensional sign they'd learned to associate with food.

I was surprised that horses could transfer what they'd learned from an image to real life. I would have thought such a thing had to be taught, and that the teaching would require a great deal of time.

In the next phase of her experiment, Stone took photos of two human faces and put them on two-dimensional boards. The horses learned to associate food rewards with one face, but not with the other. Then she released the horses into a paddock. The horses approached the person who had the face that they'd learned to associate on the two-dimensional board with food.

They were quite successful at this task, in all cases but one. For some horses, Stone had used photographs of two identical twins. The horses learned to differentiate the *photos* of the twins, but in a real-life situation, they could not tell one twin from the other.

"I can't either, most of the time," Stone told me.

I wondered how the horses would have been able to differentiate between even the photos of the twins. When I studied the faces of the twins in the paper Stone published about her research, I, too, could find slight differences in the photographed faces of the identical twins. But I had to pay close attention. I suspect that in real life I also would have had trouble telling the twins apart.

What parts of the human face were the horses looking at? The last decade of research into our own neural networks has shown that the human brain is littered with groups of neurons that specialize in processing information derived from the faces of others. At least one of these patches of cells is directly connected to the amygdala, which influences our emotional responses to the world around us, according to Kandel. This is why we respond so quickly to emotional expressions.

Do horses have such neurons? "Very likely," the neurobiologist Hans Hofmann told me, "since all mammals thus far have very similar makeup of their social brains." If so, are any of those groups of neurons in horses connected to the amygdala, as they are with us? That research has yet to be done.

Of course, research results must be able to be replicated by other scientists, using different study subjects. British researchers Jennifer Wathan and Karen McComb have also looked at how horses respond to photographs. This time, they tested seventy-two horses in order to try to understand what would happen if the test subjects looked not at human faces but at the faces of other horses placed on a two-dimensional board.

First the researchers took a photo of the head of a horse. The horse was paying attention to something that could not be seen in the photo. His eyes and ears were focused on that unseen object. The ears were pointed toward that object, and the horse's gaze was clearly directed toward that same focal point. They hung this photo on a wall, halfway between two grain buckets placed on the ground. The horse in the photograph appeared to be paying attention to one of these two buckets.

Next, they led a test horse to a spot several feet away from the photo and the two grain buckets. The horse was allowed to approach the photo and grain buckets. As he walked forward, the test horse often stopped and looked at the photo of the horse. Then he usually chose the direction in which the photographed horse seemed to be looking and put his nose in the grain bucket that seemed to be drawing the attention of the photographed horse. This showed that the test horses were taking cues from the horse in the photograph.

But what cues were these horses looking for in the photograph? In the next phase of the study, researchers blindfolded a horse and then photographed him. They also photographed a horse who had had his ears covered with a cloth. When these photos were placed on the wall between the two grain buckets, the study-subject horse wavered when approaching the photo and the grain buckets. He seemed uncertain. His choice of which grain bucket to approach was inconsistent. It didn't matter whether the photographed horse had either his eyes or his ears covered; the test horse seemed to need to see the whole face of the horse in order to make a decision.

So the research showed that horses rely on cues provided by both the eyes and the ears of other horses. Like the studies of Timney and Stone, it shows that horses can understand two dimensions and apply that understanding to the world around them, but it also confirms that horses are highly social and communicate with each other on a routine basis. They read body language and make decisions based on what they've seen. Clever Hans had developed this skill to a very high degree.

Once a horse has learned something, how long does he remember what he's learned? The French researcher Carol Sankey has shown that horses remember for long periods of time, and that early experiences may stay with them for quite a while. She has also shown that positive reinforcement is much more effective than negative reinforcement.

Sankey studied two groups of foals who had had minimal contact with humans. One group of foals was taught to stand still on command

and given a food reward when they stood quietly. The other group of foals was taught to stand still but did not receive a food reward. Sankey found that the foals who received the food reward learned more quickly and remembered their training better than the foals who did not receive food rewards. She also found that the foals' positive experiences transformed into a more positive attitude toward humans in general. "Horses are no different from humans," Sankey wrote. "They behave, learn and memorize better when learning is associated with a positive situation."

One reason why both horses and humans learn better in positive situations is because placental mammals have social-behavior and reward-behavior neural circuitry with evolutionary roots stretching all the way back to the pre-dinosaur era, suggests Hans Hofmann. When these reward circuits are activated, that memory stays with them.

The discovery of this reward circuit dates back to the 1950s, when researchers found that if rats could press a lever that delivered a small amount of electricity to a certain part of their brains, they would then press the lever over and over again, even to the point of starving to death rather than taking time off to eat. In other words, they became addicted. This center—the nucleus accumbens—is present in all mammals and is involved in many addictions, including human addiction to drugs and alcohol.

The nucleus accumbens is only one part of the reward circuit. Hofmann has studied a second circuit that's similar, the social behavior circuit. He believes this brain circuitry also evolved long ago. "These social circuits are there even in fish, which also have complex social groups. The circuits have been around a long, long time, although they probably don't function now the way they did 500 million years ago," he told me. Hofmann has found that even fish experience some level of bonding and that because of this ancient social network they sometimes even exhibit parental care.

"When you think about it," Hofmann said, "this isn't so surprising. All animals face similar challenges and similar opportunities. They have to find mates, find food, defend themselves."

Then he asked *me* a question: "Do you think horses are able to bond with humans in ways similar to dogs?"

I considered my answer. Before I started researching this book, I would have said no. Left to their own devices, I would have said, horses would prefer not to be bothered by us.

Now, I said to Hofmann, I was rethinking that point of view.

"I think they probably are like dogs," Hofmann said. "I'm not a horse person, but I'm an animal behaviorist. I do know that horses form very complex social groups and that domestication of horses wouldn't have been so successful if horses weren't able to have so much social cognition."

The idea that I could have a relationship with a horse somewhat on the level of my relationship with my border collie was beginning to dawn on me. Horses live social lives that are every bit as complex and compelling as the social lives of elephants. I thought about High Tail and her chosen stallion, and about Karen and Lukas, and realized that we have, since the days of the Pleistocene artists, neglected to fully examine the true nature of horses.

It stands to reason that animals able to form bonds can engage with and learn from each other. But what is the best learning environment for a horse? Sankey's suggestion that horses learn best through positive reinforcement has been backed up by preliminary research from the behavioral biologist Kathleen Morgan. She wanted to know if, when training miniature horses, a negative (aversive) action was more or less effective than a positive action. She studied two groups of horses: one trained to step sideways by being pushed sideways (negative action), and the other trained to step sideways by being rewarded when the correct behavior was performed. Sankey found that both groups learned the task at about the same speed, but that the horses who received the reward for the correct behavior were more willing to engage with the trainer to learn new tasks later on. Just like Lukas, the miniature horses seemed to enjoy *engaging with* their humans.

Reward rather than force set the horses up for better long-term learning. "The partnership is different when the animal is being only positively reinforced," she told me. "We got a better-quality buy-in. It was like putting money in the trust bank: if you have a lot in that trust bank, then when you have to do something that the horse really doesn't like, like floating teeth,* it might be easier to get a withdrawal from the bank than if you're already running in the red. We're investing in the relationship this way because it will pay off later."

Like Kathleen Morgan, the German ethologist Konstanze Krüger has studied how horses learn, but in Krüger's case, she has studied natural learning in horses by watching their behavior with each other. Krüger looked at two separate populations of horses—horses stabled in barns and horses allowed free range.

Earlier research seemed to show that horses did not learn by watching each other. This kind of learning is called "social learning." Krüger thought this had to be wrong. Certainly horses capable of forming strong bonds and of reading body language ought to be able to learn by watching. We know through many studies, including those of Jane Goodall, that African primates learn by watching and thus have what some call "culture"—the transmission of knowledge through generations. A generous body of research shows that this is true of nonmammalian vertebrates like birds, too.

So it made sense to Krüger that horses would also do this. A band of horses would have to learn from others where to find water holes and where to find the best places to graze. Anecdotal tales also suggest this. And as any horse owner knows, some horses watch people open stall doors and then copy the behavior. I know that I didn't "teach" Whisper how to get water out of a faucet. He had to have watched me use my hand (in his mind, possibly, my hoof) to get water out of that piece of metal, and he tried it, too. Some trainers work with young horses by first exposing them to a more experienced horse doing the same thing.

These are all anecdotal experiences, of course, but Krüger put a scientific face on this question. In one experiment, she trained a horse to follow a human. Then she had another horse stand and watch the first horse follow the human. Horse Number Two learned by watching

* When horses do not wear down their teeth by constantly grazing, they cannot close their mouths properly. Technicians take care of this problem by filing—floating—the teeth for the horses.

Horse Number One and followed the same human—but *only* if Horse Number One was older than Horse Number Two and held a higher status in the horse band. If Horse Number Two, the watching horse, was socially superior to Horse Number One, she found, the observing horse paid no attention.

Age and experience counts, in groups of horses as well as in groups of humans. Other animals do this, too, she told me. "It's possible to observe something similar in baboons. These animals do not respond to alarm calls from lower animals, but if the alarm call comes from a high-ranking animal, then they're all up the trees. That's because the older, more experienced animals have more reliable information."

"It's also of evolutionary benefit to build up knowledge from one generation to the next," Krüger continued. To remove experienced animals from a band or a herd puts the whole group at risk, she said, adding that when people shoot older African elephants, they're destroying knowledge that's essential for the survival of the next generation. The same, she said, would be true for bands of horses.

Interested in learning more about how ranking affects horses in a band, Krüger decided to study conflict resolution in horses. She has found that even foals know how to ease tensions among band members by clacking their teeth together in a gesture of appeasement.

"We've also got intervention behavior in mares and in stallions," she said. "For example, when a stallion from somewhere else gets involved in aggression between two stallions. We've found that it's not always a dominant stallion that intervenes in a conflict and that it's not necessarily the case that the stallion who intervenes rises in stature."

Peacemaking among horses—a novel idea. But her story reminded me of the three-way stallion encounter Jason Ransom and I had seen in the Pryor Mountains. I explained the story of the snaking stallion who chased another stallion over the meadow until Duke appeared from below the ridgeline. All that was needed was Duke's sudden appearance with his stern posture and his beautifully arched neck for the other two stallions to stop fighting. In that case, Duke was a dominant stallion, but, Krüger told me, the peacemaker may not always be the one in charge.

"We can't put a finger on why some horses do this frequently, while others don't," she told me. "It's a fascinating behavior. Our hypothesis is that we have more-social and less-social animals, pretty much like with people. Some are willing to intervene and regulate and some are not."

Krüger's finding is revolutionary, in that stallion behavior is almost always described in simple terms of aggression and dominance. I asked her why, evolutionarily, stallions might get involved in mediating disputes.

"We are down to two explanations," she answered. "Maybe it's just that all members benefit from this aggression regulation. Or maybe it's about the protection of and building up of social bonds."

But what about numerical ability? Can horses really understand anything about numbers, or was that just wishful thinking on our part? We now know that some animals—birds and other primates, most notably—may be able to keep track of small numbers. Can horses also do this? Clever Hans wasn't really able to do math, but Karen Murdock does suggest—anecdotally—that Lukas understands the difference between such numbers as two and four—not just the visual numerals, but the quantity concepts.

The cognitive scientist Claudia Uller, interested in the origins of thinking in animals and humans, decided in 2005 to celebrate the centennial of the unmasking of Clever Hans by daring to begin to tackle the problem of horses and numbers.

"We adapted a task originally devised to be done with infants," she told me, "so that we could see whether horses were as sensitive to numbers as human babies. Babies are not only perceiving these things, but are keeping these representations in their memory. Others have claimed the same kinds of things for other animals. That's the theoretical idea for the horse experiments—we wanted to show this for horses. To our knowledge, it was the first time that we were actually showing numerical representations in horses."

Uller and a graduate student, Jennifer Lewis, showed horses two plastic apples and three plastic apples. The apples were plastic so that researchers could be sure that horses weren't using their noses to solve the problem. (We assume that the horse's ability to detect smells is better than our own, but as the French neuroscientist Michel-Antoine Leblanc explains in *The Mind of the Horse*, scientific research in this area is sparse.) The horses were allowed to watch while researchers put the group of two apples in one bucket. They placed the group of three apples in a second bucket. Both buckets were opaque, so the horses could not see the apples once they were placed inside. To get the apples, the horses had to *remember* which bucket held the greater number of apples.

Uller and Lewis found that most horses approached the bucket with three apples rather than the bucket with two apples. It's important to understand that this experiment was not an experiment that showed that horses could *learn* the difference between the two numbers. Horses had to understand *innately* the difference between two and three.

"You're testing an ability that is spontaneously available. There is no training at all. No learning. Each horse gets only one test, one trial. The important thing about this experiment is that the horses had to keep the two numbers in their memory. It's a quite sophisticated thinking process," Uller said. "This is particularly interesting because 'number' is an abstract ability. It's not like knowing the difference between an apple and an orange. Number is in the mind: The 'threeness' of a set of apples. You build a representation in your mind of 'threeness.' This shows that horses can think abstractly, and they do not depend on language to do so."

This is just a preliminary study, which needs to be confirmed, but it falls into line with other recent studies that show that our ability to understand and communicate with animals is based in part on abilities that may have evolved quite early in the history of life.

Beyond the details of gross brain anatomy, we don't know much about the similarities and differences between human and horse brains, but we are learning a few facts about the similarities between the brains of dogs and humans, which tells us something about why we are able to work together so well. The Hungarian researcher Attila Andics and his colleagues taught eleven dogs to lie still in an MRI scanner. Researchers watched which parts of the dogs' brains responded when the dogs heard other dogs bark, and which parts responded when the dogs heard people talk. Then they put humans in an MRI and ran the experiment in reverse, watching which parts of the human brain responded when they heard other people talking and when they heard dogs barking.

They found that certain voice-sensitive areas in the brain lit up in dogs as well as in humans when they heard communication sounds. Dogs responded more strongly to the sounds made by other dogs, but they also responded strongly to human speech. Humans responded most strongly to the sounds made by other humans, but also responded to the sounds made by dogs. Researchers suggest that those voice-sensitive areas are also likely present in the brains of other mammals.

Horses have yet to be studied this way, but I'm holding out hope. It would be fascinating to learn the details of what might be going on in horses' brains when we talk to them. Most horse owners will tell you that horses respond to the sounds other horses make and to the sounds made by human voices, and also that horses can tell one human voice from another. But that's just barn talk.

Now the British researcher Leanne Proops and her colleagues have confirmed in behavioral experiments that horses do indeed connect the "voices" of individual horses with the horses themselves. She led one member of a band of horses up to and then away from a test horse. When the horse being led could no longer be seen by the test horse, she played a recording of the voice of an unfamiliar horse. The test horse paid close attention to the unexpected call of the strange horse, and less attention to the sound of the familiar horse who had just passed by. The call of the strange horse violated the expectations of the test horse, Proops wrote. Later, she found that horses recognized and responded to their owners' voices, but not to the voices of strangers.

To horse owners, the results of these studies might seem so obvious as to make them not worth doing. But as I've mentioned, the scientific study of the thinking abilities of horses has been off-limits for quite a long time. Because of the Clever Hans stigma, we are only now beginning at the beginning. Each of these small studies is helping to build a foundation for further work. And I suspect that some of this research may reveal that horses spend time with us not just because we have the food and the water—but because, sometimes, they just want to be with us.

I think they enjoy the companionship—the partnership works both ways.

When I was in Vienna attending the wild-horse conference where I met Jason Ransom and Laura Lagos, I visited the Spanish Riding School, where the famous Lipizzan horses have been trained for hundreds of years. It was there that I interviewed Herwig Radnetter, one of the school's riders.

I've always loved these horses. As a small child, I was fortunate enough to see them perform in Madison Square Garden, and later I was allowed to "ride" one. (Well . . . actually . . . to sit while someone else led the horse around.) Lipizzans originated in Spain and were introduced to Vienna by the House of Hapsburg. For hundreds of years, they have been pampered like princes. The bedding in their stalls is clean enough to sleep on (at least, I would), and the horses have their own personal grooms who check on them constantly.

Today they hold the status of a living European art treasure, as revered as the Parthenon or the *Mona Lisa.* At the end of World War II, though, the Nazis and the Soviets nearly destroyed the breed. The horses were only saved because of the heroism of Alois Podhajsky, then in charge of caring for the horses, and of several other Viennese and German men who willingly put their lives in danger to get the stallions out of Vienna before they were killed.

The horses also owe their survival to the bullheaded determination of General George S. Patton, then commander of the Third U.S. Army and a devoted equestrian. Patton ordered Colonel Charles Hancock Reed and his men to rescue the horses in an exercise that came to be known as "Operation Cowboy." A fictionalized version of this tale can be seen in the 1963 film *Miracle of the White Stallions.*

There has always been a sense of magic attached to these ethereal animals, but the Disney movie increased their glamour. The first time I saw the horses in Vienna, a long line of them were being led, groom by groom, across a public street from their stables to their performing arena. The crowd on the street stopped, awestruck. One woman reached out and touched a stallion. The horse kicked out, and a groom severely reprimanded the woman. She should not have done it. But I understand why she did. Their liquid eyes are irresistible. The stallions are so otherworldly that it seems as though if you touch them, they'll disappear.

During my interview with him, I asked Radnetter about his involvement with the horses. He explained that when a new, young rider begins his education at the riding school, he is matched with his own set of three young horses. The riders and their horses then undergo a grueling training routine that may last for four or five years. Many of the riders who begin the program drop out after a year or so. The training and discipline are rigorous.

Radnetter stressed to me the importance of patience—the same lesson stressed by Kris Kokal and by Karen Murdock. In particular, he said, it's essential to form a strong bond that's never breached. The horse cannot lose faith in his rider. It's this trust that makes the stallions amenable to the exercises they undergo.

Radnetter was not the first Lipizzan trainer to talk about the importance of this bond. In his 1965 book *My Dancing White Horses*, Alois Podhajsky discussed how strong the connection can be. Once when he was riding his horse along the Danube, the animal stumbled and fell into the swiftly flowing river. Podhajsky, who jumped off rather than go in with him, stood on the riverbank and watched his horse float away. The horse didn't seem to know what to do. Without much hope, Podhajsky called the stallion's name. "My voice worked wonders," he wrote. "Bengali lifted his head, neighed feebly, and struggled mightily against the current in an effort to reach me on the bank." The successful effort left Podhajsky rejoicing in the strength of the bond.

Later, Radnetter introduced me to his own three stallions. The moment they heard their rider's voice, two of the stallions, stabled next to each other, lifted their heads from their hay and poked their noses out of their stall doors. At their response, Radnetter beamed.

Next we visited his other stallion, stabled farther away, off by himself, in a quiet place out of the limelight.

"This is my 'autistic' horse," he told me. He had a smile on his face.

I asked him what he meant.

"When we go away from here to perform anywhere, he tries to hide by putting his head under my jacket. He doesn't like strangers or new experiences and he can get quite upset."

"Why do you keep him, then?" I asked.

Radnetter looked shocked.

"Well," he said, "he's my horse."

10

THE REWILDING

> Only the wind can ride you now.
>
> —BAGI, hero of the Mongolian film *Khadak*, to his companion horse as he returns him forever to the Mongolian plains

What do we owe to these horses who have traveled across time with us, who have carried us across the North American plains and the Asian steppes, who have plowed our fields and helped feed us, and who have provided us with profoundly moving aesthetic pleasure? As our own human population grows, our world seems to be getting smaller. Now that we don't need them any more for their horsepower, do horses still have a place in our lives? What place should that be? And how can we pay them back for everything they've given us?

To find out more about the future of the horse, I traveled to Mongolia, where horses reign supreme. Most horse people I know care deeply about their animals, regardless of their particular style of horsemanship. The first person who ever told me to be kind to a horse was a grizzled cowboy who'd spent so much time in the saddle that he had to wear a hernia belt. "Be nice to this horse," he told me, while saddling my mount and looking up at the steep Rocky Mountain cliffs where we were about to ride. "He's going to keep you alive. He's your friend. You need him up there." Every once in a while, though, this caring reaches genuinely heroic levels.

Long before I began thinking about writing this book, I read about the rewilding of the Przewalski's horse and was deeply moved by the decades-long dedication of a few of the volunteers who had made this happen, including a key Dutch couple, Inge and Jan Bouman. I had to go to Amsterdam for a scientific conference several years ago and took the opportunity to invite Inge to lunch in Rotterdam, near her country home.

We met in a busy café. It was a thoroughly twenty-first-century restaurant in a thoroughly modern city that's been almost completely rebuilt after the destruction of World War II. In the United States, that war seems like ancient history, but there are still plenty of Europeans alive who, when very young, experienced this nightmare firsthand. In Europe, people talk about the agony of this conflagration as though it happened yesterday. The cities may have been rebuilt, but the memories live on.

By the time I met Inge for lunch, I had become somewhat accustomed to this, but I wasn't prepared for the tale she was about to tell me. It would turn out that the motivation for Inge and her husband, Jan, to rewild the horses had a great deal to do with their dreadful experiences during the war. In the late 1930s, Inge's parents were working in Malaysia for a Dutch company. When the war broke out, the family was interned in a Japanese prison camp. Inge passed much of her childhood staring through barbed-wire fences. After the war, she returned to the Netherlands but never felt she belonged. Her culture was the culture of prison camp survivors, rather than of any particular nationality. Children from all around the world had been her friends. As an adult, she became a child psychologist, dedicating herself to the care of abused and neglected children.

Jan, Inge told me, was the son of a wealthy businessman and had lived out his prewar childhood in the Netherlands under the despotic reign of an unhappy stepmother. At night, he managed to escape the woman's ire by sleeping in the barn. His pony was his comforter and companion. The equine made the boy feel so much less alone and frightened. For the rest of his life, ponies would hold a special place in his heart.

By the 1940s, Bouman had become a businessman, like his father before him. When the Nazis marched into Amsterdam, he experienced the misery and fear of the war firsthand, as did all city residents. Additionally, he was responsible for the care and protection of many employees and their families.

Fast forward to well beyond the end of the war, when his special bond with ponies and his protective nature would combine. Most of Europe's zoos had been devastated. Animals were either abandoned in their cages to die or let loose to fend for themselves in bombed-out cities, often to be eaten by starving people. Zoo populations were severely reduced. Among the species almost eradicated was the Przewalski's horse, called by Mongolians the Takhi horse.

The Takhi is an odd, stocky equine with short legs, an oversize head, a huge jaw, and a stubby zebra-like mane. He has the poorest excuse for a horse's tail that I've ever seen. In France, many of the guides said that the Takhi looks like some of the horses painted on Pleistocene cave walls, but there's no evidence that these horses lived in Europe at that time. The Takhi was "discovered" in Asia by Europeans at the end of the nineteenth century (the Mongolians have always known about them), and there was a subsequent rush to collect them for European zoos. Hundreds were caught for shipment to Europe, but most died during the perilous trek west. Of those who survived, many could not adapt to confinement in tiny zoo paddocks. And of those who could adapt, many disappeared at the end of World War II. By the 1960s, few Takhi remained alive, and only nine zoo animals were able to reproduce. This extremely limited gene pool resulted in precariously low reproductive rates.

Even worse, in Mongolia all the free-roaming Takhi had become extinct. The cause of this extinction, like that of the extinction of horses from the Western Hemisphere, remains elusive, but it may be somehow connected to the increased number of people living in Mongolia, or to changes in the Mongolians' traditional lifestyle wrought by Soviet mandates and bureaucracy. Mongolians have traditionally lived highly nomadic lives, which allowed them to respond to climate anomalies simply by moving camp. The Soviets, however, insisted that the Mongolian people settle down and live permanently in collectives, which inevitably stressed the delicate steppe lands.

Aridity increased, as did the number of domestic animals—goats, sheep, cattle, domestic horses—competing for grazing opportunities. It may be that these domestic herds pushed the Takhi over the edge. Or, since domestic horses and Takhi can interbreed, it may be that the Takhi simply interbred with other horses and disappeared as a separate species. It's also possible that European zoo collectors had ravaged the wild population to such an extent that free-roaming bands became genetically compromised and could no longer produce adequate numbers to sustain the population. In any case, whatever the reason, the last wild Takhi was seen in the late 1960s.

In the decades after World War II, Jan read a great deal about the plight of these horses, and had become concerned. In his imagination, the Takhi was a noble animal, a valiant ancestor to the pony who had kept him company as a child, and in 1972, he and Inge decided to travel through Europe visiting the remaining zoo animals. They planned their trip with great excitement, but when they actually laid eyes on the horses, Jan was distraught.

These were not the wild horses of his dreams. Instead, they were forlorn and obviously depressed captives being held behind bars in tiny paddocks where they could barely move. Their situation resonated with the couple. The horses stood around with hanging heads, like retired plow horses.

"Numb," Jan and Inge said to each other.

The ground on which the listless horses stood was nothing but dirt. They had no grass to eat. They didn't bother to look at each other, let alone interact, except to kick or bite. Even this they did halfheartedly, as if establishing their own personal space wasn't worth the effort. The horses seemed to be half alive. They reminded Jan of life in Nazi-dominated Amsterdam. They reminded Inge of what it felt like to live life behind barbed-wire fences.

But at least the horses were breathing.

The couple had learned one important lesson from their wartime miseries: Where there's life, there's hope.

Jan decided to liberate them.

Jan and Inge decided to search for a way to bring the horses back to the steppes, where the animals could once again roam at will, and thus bring them back from the brink of extinction.

The process of taking animals out of captive situations and returning them to the wild—often called "rewilding"—is never simple, but the task of rewilding the Takhis was even more complex. Horses are fairly large animals who roam in extensive home territories and who, as Jason Ransom and Laura Lagos had explained, must live in tightly knit social bands in order to thrive. Unfortunately, there's just not a lot of open space available in Europe, save in forested areas. Lands that would have been suitable for roaming bands of horses had long been used as farmland.

Adding to the complexity was the political situation in Europe and Asia. Following World War II, international communication and cooperation were limited, as each nation struggled on its own to find ways for its people to survive and for its economy to revitalize. Additionally, many nations—including nations with land suitable for free-roaming horses—had been absorbed into the Soviet Union. If the Boumans were to search for the best grazing areas for wild horses, they would have to travel behind what Winston Churchill once called the Iron Curtain, which would add layers of diplomatic challenge to their already difficult task.

Nevertheless, from 1972 until Jan's death in 1996, the couple worked tirelessly to achieve their goal. They left their jobs, moved to cheaper quarters so they could spend their money on saving the horses, ignored people who told them that what they wanted to do couldn't be done, and traveled throughout Europe, North America, and Asia at their own expense, explaining the plight of the horses to whomever would listen. They compiled painstaking breeding records on handwritten cards of the horses kept in the zoos, so they could avoid—wherever possible—unfortunate genetic mismatches that would result in unviable pregnancies.

They began their mission independently, without financial support from zoos or governments or wealthy organizations. They raised funds for their project in bits and pieces by selling place mats and T-shirts at fairs and bake sales. At one conference, Jan spoke out with so much conviction that he ended up hospitalized with a heart attack. Their efforts were not always welcomed by European zoo officials, who felt the couple lacked credentials and status. In the then-more-freewheeling United States, the couple was well received, but help was nevertheless not forthcoming, because Americans believed that, for the most part, the rewilding of this species was a European endeavor. At times, Inge told me, they felt completely hopeless.

But they persevered. With the help of many other volunteers, Jan and Inge formed the Netherlands-based Foundation for the Preservation and Protection of the Przewalski's Horse. Eventually their emotional commitment spread nationwide. The cause of the Przewalski's horse became a Dutch obsession. News articles were written about the horses and about the Boumans' determination. Government officials began speaking out in support of their efforts. The word "Przewalski's" was even used in a Dutch national spelling bee.

An international captive breeding program, a joint effort that coordinated zoos from all over Europe and the United States, had managed to increase the total world population of Takhi horses to well over five hundred animals. This was an improvement, but the number was still by no means adequate.

As the population of Takhi increased, Jan and Inge and others began to think about how to get the horses out of the zoos and onto grazing areas, which Jan and Inge called "semi-reserves." In these pastures the horses would have at least some freedom.

However, Jan and Inge hoped for an even better life for their charges. They envisioned a location for the Takhi that would allow them to roam over large ranges, to choose where to graze, where to find clear-running streams, and where to live as they had evolved to live. For that they needed a lot of open space. The best place, they thought, would be the Asian steppes. The grasslands of Ukraine, for example, would have been perfect. Unfortunately, those grasslands and other, similar regions were controlled until the 1990s almost entirely by the Soviet Union. Sending the horses there didn't at first seem possible.

Then, in 1998, in the midst of Mikhail Gorbachev's openness policy, Soviet officials contacted the couple and suggested several possible locations in Siberia, Ukraine, East Kazakhstan, and Mongolia. They invited the Boumans to visit these sites and to help decide which would be most suitable as a permanent home for the horses.

Unfortunately, after touring the areas, the couple turned down each of the suggested sites. Some were too small. Some had limited grazing opportunities. Some would have required the eviction of people living on the land, which the Boumans were unwilling to do. Some were already used for grazing domestic horses and other animals. The last thing the Boumans wanted was to try to rewild the Takhi in a place where the local people did not welcome the horses.

Frustrated and in despair, Jan and Inge wondered if the world no longer had room for the wild horses. But then, just as they had reached their nadir, political conditions changed. Following the dissolution of the Soviet Union, Mongolia became an independent nation. Mongolians were finally able to once again celebrate their own millennia-old local culture, and Jan and Inge found a nation that was yearning for the return of their fabled wild horse.

The Dutch couple had found a perfect trifecta: open space with good grazing, a welcoming culture, and suitable political conditions that allowed the world's scientists and other interested parties the freedom to easily come and go in order to work. Indeed, Jan and Inge found that in Mongolia, the Przewalski's horse was not only welcomed, but revered. Mongolians still sang old songs that memorialized the Takhi's strength, resilience, and determination. Poets praised the wild horses. Traditional tales dating back hundreds of years extolled their bravery. The people of Mongolia had never seen living Takhi horses, but they remembered them.

This cultural memory was central to the success of the rewilding project. One of the hardest tasks in rewilding any animal is to get the full support of local people. Without that support, a rewilding project is doomed to failure. One by one, the animals will disappear, ending up on someone's dinner plate, caught and sold to black-market buyers, or just killed and buried where they'll never be found. This is particularly true if the region is poor and the people need money and food. It's difficult to tell a father with hungry children that he should not hunt wildlife or that he should allow wild animals to eat the grass that his own domestic stock needs. Without human support, the rewilded animals cannot survive.

But here in Mongolia that problem was not as acute. Because the Takhi were so loved, many people were not only willing to tolerate the animals, but enthusiastic about their return. Unfortunately, even in Mongolia, finding room for the horses proved difficult. The land is not fenced and looks to Western eyes as though it's wide open, but in fact the country has millennia-old customs of how the land is to be used and by which families. Mongolian nomads don't just "wander" with their livestock any more than Pleistocene people just "wandered" over the European landscape. Many Mongolian families travel to various grazing locations on a season-by-season basis and often have traditional rights to use these places. While the land isn't "owned" in our Western sense of the word, these traditional rights are still important. If the horses were to be placed somewhere, the people using the land had to agree to alter their patterns of use of the area where the horses would live.

This posed a difficult problem. Then, someone suggested that they visit a region located about an hour's drive west of the Mongolian capital, Ulaanbaatar. Ecologically, the area seemed perfect. A river full of water year-round ran through plentiful grasslands. Near the river were mountains, ravines, and trees, where the horses could shelter from the summertime heat or the wintertime wind and snow.

But best of all, at the center of the site rose Hustai Mountain, a sacred Buddhist refuge where people had come for hundreds of years to pray and to celebrate. Protecting the sanctity of this mountain was a goal that many Mongolians would support, so setting aside a large parcel of land with the sacred mountain as the centerpiece would be a popular idea.

On the other hand, there were many herders who grazed their livestock in the valley regions surrounding the mountain. The site also had a short valley passageway that had been used by nomads as a throughway for as long as anyone could remember. Cutting off access to that valley by creating a park would be a serious breach of tradition. Jan and Inge, with the help of Mongolian scientists and supportive officials, negotiated agreements that allowed for traditional uses and also protected the horses. Ultimately, all parties agreed that returning the Takhi to Hustai would be a win-win situation for everyone. Since the project would be extremely expensive, the Dutch government even agreed to help with funding.

Then, finally, on June 5, 1992, the first Takhi horses returned from Europe to their native land.

"You should come to Mongolia next month," Inge told me at our lunch, after telling me this story. "It's twenty years now since the horses came home. We're having a celebration."

And so I did.

If the Takhi horse belongs to any one region, it's the world of inner Asia, where the land is still not fenced and where the rolling hills look like the rolling ocean and where in most of the country nothing can be seen but grasslands and bush and where walking is so discouraging that you think you'll never get to the next rise, but you know you could gallop on forever if only you had a horse; and where dawn horses and *Hipparion* mares and eventually *Equus* himself once roamed with abandon, grazing on the green grasses and fields of flowers brought to humans and horses compliments of the 100-million-year-old Cretaceous Terrestrial Revolution.

This is Horse Country, writ bold.

All you have to do is look at the Takhi to know that he belongs here. But you'll also recognize that he's different from the horses we're used to seeing. He lacks the elegance of the Vogelherd horse. A dark stripe running down his dun-colored back bespeaks his heritage. His front nipping teeth are quite large, much larger than in other horses, and his shockingly huge jaw is more than beefy enough to house his massive molars.

He is also unridable. By nature, he is intransigent. The stallions are sometimes extremely aggressive. The Smithsonian Conservation Biology Institute has a Takhi breeding program based in Front Royal, Virginia, where veterinarians are learning how to artificially inseminate

A Takhi stallion in Mongolia's Hustai National Park (Greg Auger)

the horses, an effort which will help in increasing the global population. While visiting the Smithsonian's Front Royal facility, I stopped by the town's visitor center, where I chatted with a volunteer who told me that he had once been standing casually by a fence enclosing a Takhi stallion. "Nice horsey," he was thinking to himself, when, in an instant, the stallion rushed him with teeth bared, prepared to take a chunk of human flesh. The volunteer was impressed, but not overly enamored.

The Takhi is not easy to get along with, true, but the Mongolians don't mind. A national symbol, he represents the Mongolians' spirit of independence. Mongolians believe it's a wonderful thing to have a foal from a domestic mare sired by a Takhi stallion, because his genes provide toughness and stamina. The Takhi is a genetic anomaly. Riding horses, including Mongolian riding horses, have sixty-four chromosomes, but the Takhi has sixty-six chromosomes. Nevertheless, a Takhi and a domestic horse breed easily and can produce a fertile foal. This foal has sixty-five chromosomes, but if the foal breeds again with a domestic horse, the offspring will be a horse with the standard sixty-four chromosomes.

Some researchers once suggested that Takhi horses were the ancestors of modern horses, but recent genetic research has found that the Takhi separated from the main trunk of horse evolution about 160,000 years ago, only shortly after *Homo sapiens* evolved on the African plains. So the Takhi is a side branch in horse evolution, perhaps a modern version of *Megahippus*, the evolutionary dead end that Matthew Mihlbachler discussed at the American Museum of Natural History. They are not the root from which modern horses grew, but they are close kin.

That the Takhi is an emblem of modern Mongolia became obvious to me the moment I stepped off the plane in Ulaanbaatar. Mongolians adore this animal with a fervency I had not expected. I spoke with a number of people about them, and everyone was thrilled to have them back. Of course, Mongolia is all about horses anyway. The emblem of the Mongolian state airlines on which I flew is a blue horse's head encircled in yellow, the blue signifying that Mongolia is the Land of the Eternal Blue Sky and the yellow signifying that the sun always shines over this landlocked nation. Fixated as Mongolians are on horses to begin with, they have a special place in their hearts for this audacious, untamable, prodigal horse that was lost to them and has now come home.

Each time I said "Takhi" people melted into paroxysms of enthusiasm. They often mentioned that the return of the horses occurred at about the same time as their nation gained independence from the Soviet Union, doubling their love for the horse. As the Liberty Bell is to the American spirit of independence, the Takhi is to the Mongolian spirit of freedom.

Ulaanbaatar, two decades ago a small country town where horses and red deer were more common than cars, is now a major international city. Tracking down the location for the celebration was not an easy task, but my taxi driver persevered. It turned out I had been given an incorrect address: the location was a back alley. Next he took me to a concession center for Western hunters looking to bag Mongolian big game. The hunting guides very quickly figured out that I didn't belong there.

They sent us over to a Buddhist monastery, where the golden-robed monks had no idea what we were talking about but smiled benevolently anyway. I waited for the driver to throw up his hands in despair, but he wasn't giving up. Not for nothing have these purposeful people survived millennia of being relentlessly squeezed by China to the south and Russia to the north, and not for nothing did Genghis Khan establish an empire ruled from horseback that extended from Korea to the gates of Vienna.

Finally, we pulled up to a small nature center. Photos of Takhi covered the walls inside. My driver looked triumphant. The place was filled to overflowing with Mongolian people and a few Westerners, including Inge and her colleague Piet Wit, a prime point man for the Takhi effort and an urbane wildlife expert with a lot of experience. Aside from helping to organize Hustai's Takhi reintroduction program, Wit has worked extensively with African programs that train local people in protecting wildlife, managing water resources, and implementing responsible forestry practices. In the politically complex West African nation of Guinea-Bissau he established, in honor of his son who died tragically at a young age, an important research and protection program, called Chimbo, to protect the highly endangered West African chimpanzee. He is also chair of the IUCN Commission on Ecosystem Management. Rewilding requires considerable on-the-ground experience, and Wit, who has committed his entire life to this effort, is one of the world's best.

By the time I walked into that celebration, the Takhi had been in the country for two decades, but it was clear from the vodka-filled three days, after listening to speech after speech praising the return of "our dear horse," that no one was yet taking them for granted. The horses held the status of a living treasure, like the Lipizzans of Austria. Although I tried, I learned very little of the Mongolian language during those three riotous days, but I did often hear, about every ten minutes, someone saying proudly, "Mongol Takh, Mongol Takh." It's a phrase I will never forget.

During this time, I met many people who participated in the effort to preserve the prodigal ponies. After the first horses were returned to Hustai, many more Mongolian officials and scientists joined the effort. Jachin Tserendeleg oversaw the breeding of the horses. He died at about the same time Jan Bouman died, but his son is still involved. And I met Bandi Namkhai, the director of Hustai since the project's earliest days. The Hustai project began as a way to provide a protected area for the horses, but it has since grown enormously. Hustai itself is now designated a national park, and today it protects not just Takhi horses, but threatened bird species, gazelles, red deer, marmots, wolves, badgers, large and small cat species, and numerous plants.

Even Argali sheep have moved into the park, which provides a refuge from hunting pressure, something of which Wit and his Mongolian colleagues are particularly proud. Today, the park, started for the horses, has become an important wildlife haven, as often happens in rewilding projects. In Mongolia and throughout central Asia, following the collapse of communism, economic chaos ruled. Governments were unstable, and people had no experience with trade or markets, which had been outlawed for decades. The chaos and lack of dependable trading systems caused starvation and extreme poverty. Under those circumstances, the desperate people decimated wildlife populations.

Wit and Bandi Namkhai assumed the unenviable job of convincing local villagers to respect the park's boundaries when hunting. It wasn't easy. It was one thing to convince people to support the presence of the Takhi and quite another thing to convince them not to hunt the other animals and not to sneak their herds of goats and sheep and domestic horses onto the park's rich grasslands when their own grazing areas were depleted. First Jan and Inge, and then Bandhi and Wit, worked hard to convince local people that their willingness to comply would result in the long run in a healthy tourism industry that would support the area's villagers.

Today the park has become an economic mainstay for people who live in the area. Thousands of people come from all around the world each year to see the horses, to sleep in simple, well-maintained *ger*—small round portable houses with beds and stoves for wintertime use—and to tour the park, either on foot, in guided vehicles, or on horseback. Additionally, the park's foundation has initiated a microcredit program that allows individual villagers to borrow small amounts of money in order to start up small business endeavors. Wit explained that the loan program had worked quite well.*

All this was explained to me at the celebration in the city, at which I also met officials with the International Takhi-Group, a Swiss-based organization. The ITG operates a second rewilding effort in the Gobi Desert, on the border with China.

This international effort has borne fruit. By the time of the celebration, almost two thousand Takhi lived in a mix of zoos, pastures, and wild places worldwide. When Inge and Jan began their crusade to save the horses, wildlife researchers had very little experience with how best to return captive animals to the wild, but by the time of the Mongolian celebration, it had become widely accepted that zoo animals cannot be sent directly to wilderness areas without some support in the initial stages.

As Ransom and I had discussed, wild animals require extensive knowledge of their environment. Some natural horse behavior may be inborn, but the knowledge of how to survive in a region can only come from direct experience. Having done nothing but stand in small, dusty paddocks, zoo-bound animals don't even know that they *need* to know things. They don't know that they have to search for food and water, since it's always been brought to them. They don't know what kind of social structure they should adopt, or how to depend on each other, or even which species are dangerous predators and which they can ignore. They've been so isolated that they have no idea how to survive.

Since the Takhi did not know how to form such groups, Jan and

* More than 90 percent of these loans are successfully repaid.

Inge wondered if they would ever be able to survive on their own. First, the couple purchased a few surplus horses from zoos and released them into a small pasture not far from Amsterdam. Caretakers watched over them while the horses learned how to be horses. First, mare bands began to coalesce. These mare bands began to learn where to find food and water, and eventually well-worn paths appeared in the pasturelands as the horses walked from grazing to water to grazing and back to water again. In these semi-reserves, the horses began to expand their repertoire of behavior. With each passing month, they began to look more and more like other free-roaming horse populations.

Scientists saw unexpected results. Some of the effects of domestication began to disappear. For example, Takhi reproduction rates improved dramatically under these more natural circumstances. In the reserves, reproductive rates nearly tripled, increasing to about 92 percent from the 35 percent of horses kept in zoos. One reason why this happened may be that the mares' breeding dates and delivery dates became more attuned to the rhythms of nature. In domestication, mares are often bred to deliver in January or February. Zoo managers had timed the Takhi schedules this way. When mares from the zoos were first released on the reserves, this situation continued—but only for the first year. By the second year, the mares began to mate at the right time, during early summer, when the sun was high in the sky. They then delivered almost a year later, in late spring. Foal survival rates improved. The warmer weather and better forage improved their health and the health of their dams.

The strategy of releasing Takhi horses into semi-protected areas was so successful that it is now accepted practice. Today Takhi live on reserves, where they are closely watched, in a surprising number of places—in Holland, France, Britain, and elsewhere in Europe; in Ohio, in China, and even in Chernobyl, where many species of animals have been released onto land that's had to be vacated by humans after the nuclear disaster there. Some of these horses in reserves may be sent on to wildlife refuges like Hustai, but others will likely spend their whole lives on the reserves. There are still few places in the world where the horses live truly wild lives, but their populations in the managed reserves are quite healthy.

This success alone was reward to Inge and Jan for their lives well lived. During the festivities, Inge and others held a press conference to talk to Mongolian reporters about the project's success.

"It's a dream come true," Inge told them. "My husband would never have dreamed there would be so many horses."

Then she cried.

The day after the press conference, the local English-language newspaper, the *UB Post*, carried a huge headline splashed across the front page above the fold: "Mongolia Has the Largest Population of Wild Horses in the World." A photo of six mares and a foal grazing in Hustai, set against a backdrop of the eternal blue sky, spread across all six columns.

"An important reason why this project has succeeded," Piet Wit once told me, "is because of the tremendous enthusiasm of the Mongolian people."

When I drove out to Hustai, where the celebration continued, I saw what he meant. Local villagers had gathered for their own festivities. The area of the park where the tourists stayed was filled with villagers and visitors dancing and singing and eating. A lot of vodka was passed around. On a nearby hillside, Mongolian men took part in wrestling activities.

While the wrestling went on, a horse race began. Mongolian horse racing isn't like horse racing in western cultures. For one thing, some races last for hours. The hardy little horses are ridden by the smallest children in a family, boys or girls, because they weigh the least. As soon as the kids are old enough to ride by themselves over long distances, they are allowed to race their family's horses.

As I watched, parents threw their kids up onto the horses' backs. Readying themselves for the competition, the children began to ride in circles and sing traditional songs to encourage the horse to run a brave, gallant race. Eventually, the kids and horses gathered at the edge of a long, open, flat grassland. This was where the race started. Adults on motorcycles pulled up. These adults, it turned out, were going to ride along with the kids, to be sure no one got hurt.

And—the race was off. The kids disappeared in a cloud of dust. After a few minutes they were no longer visible, and everyone returned to the wrestling match, which was still going. Food appeared. People basked in the sun. The wrestling continued. The kids still hadn't returned. More wrestling happened. People talked among themselves. The kids still hadn't returned.

After nearly an hour, a few sweaty horses and riders appeared from over the horizon. Then more and more horses appeared, accompanied by adults on motorcycles. It seemed that everything had proceeded safely and that no kids had fallen or been injured. The horses were all still galloping.

As the front-runners neared the finish line (where, exactly, that was, I still hadn't figured out), riders used their whips with great enthusiasm. Mongolian horse races are free-for-all experiences, and I saw several of the riders use their quirts not on their horses, but on other riders. No one seemed upset by this.

One by one, the other horses came in. There was no great cheer for the horse who came in first, although there was polite clapping. These races are about winning, but they're also about other issues of equal importance—like the gathering of far-flung families, and the sharing of information about the year's grazing opportunities, and the worth of the individual horses owned by each family.

Horses have been central to Mongolia since recorded history began, particularly during the early thirteenth-century days of Genghis Khan. Mounted warriors ranged from Korea all the way to Vienna. This empire held together and lasted for several more centuries because Mongolian riders carried messages all over the empire, in much the same way that Pony Express riders in the nineteenth century carried messages from Missouri across the Great Plains and the Rocky Mountains to California. And just as the Pony Express riders changed horses at outposts spaced along their ride, Mongolian messengers during the era of Genghis Khan changed horses at regular outposts maintained by the empire's government.

The domestic horse is just as important to Mongolians today as it was then. It is the center that unites the nation. Mongolian herders keep a variety of livestock, including camels, yaks, goats, and sheep. But the horse is the most important. Mongolia is known as "the Land of No Fences," and the domestic horses graze freely. Cattle and other livestock are herded, but the horses are left to their own devices, although people do periodically check up on them. Nevertheless, as in Galicia, each horse is owned by a particular family. When someone wants to ride, he brings one of his horses in from the herd, halters him, and ties him to a line near the *ger*. He may ride this horse from time to time over several days, leaving him on this tether when the horse is not being used. During this time, of course, the horse is fed and watered. After a few days, when the rider thinks the horse needs to rest, he returns the horse to the family herd and gets another.

Despite this on-again, off-again existence, Mongolian riding horses are known for their loyalty to their riders. When they're grazing with their herd, they stay with their herd. But when the rider brings them in, they stay with the rider. Mongolian history and folktales are full of stories of such loyal horses. Loyalty between horse and rider went both ways: Warriors tied hairs from their favorite horses' manes onto their weapons, and "the wind in the horsehair inspired the warrior's dreams and encouraged him to pursue his own destiny," writes Jack Weatherford in his fascinating book *Genghis Khan and the Making of the Modern World*. These favored horses were often honored by their riders by having blue ribbons woven into their manes.

Genghis Khan knew very well that his success depended on horses. When Mongol warriors conquered a city, as an act of submission the armies demanded that the surrendered people pay tribute by bringing fodder for their horses. Mongolians on their hardy steeds—five for each warrior—might well have conquered all of Europe, but "where the pastures ended, the Mongols stopped," Weatherford writes. Without grass, the equivalent of oil in our modern world, they could not go any farther.

From the east coast of Asia on into the West, defenders could not stop the onslaught of the Mongols, but a change in the ecosystem could. Without the steppe conditions on which the Mongolians and their horses flourished, Genghis Khan's armies could accomplish nothing.

While it lasted, this horse-based empire created a Pax Mongolica. Business prospered. Mongolian administrators, aided in their communications with each other by their horses, worked hard to eliminate corruption from the vast interior of Asia. They even standardized a system of weights and measures. While the Khan's administrators were not lenient masters—Mongolians had no trouble slaughtering whole villages of people who were not willing to submit—the mounted warriors and government officials brought a new kind of order to a once-chaotic region.

To some Western eyes, the power and stamina of these horses who created an empire may seem surprising. Mongolian domestic horses are much smaller than Thoroughbreds, and appear to Western eyes as though they are nearly starved. Even at the end of a summer of good grazing, they have almost no fat on them. They are not glamorous. But they are strong. In a mile-long race, they don't stand a chance against Thoroughbreds, but in a race that lasts for hours, they are winners. And in a race that goes on for days, when the horses are ridden, rested, then ridden hard, over and over again, nothing can beat them. Mongolian horses aren't pretty, and they aren't schooled to perform subtle movements like the *levade*, but they are resilient, and during the time when Mongolia ruled much of the world, they were very effective. In battle after battle during the era of Genghis Khan, these little horses of the steppe consistently outperformed the much larger horses of the West.

Mongolians have long believed that one of the reasons their horses are so tough is because they have Takhi blood in them. Whether this is true or not (it hasn't been studied), this folk belief has caused some difficulties for the Takhi reintroduction. Bandi Namkhai and Piet Wit had to work hard to convince Mongolian horse owners not to sneak into Hustai and breed their domestic mares with Takhi stallions. Several times, when Piet and I drove through the park, spotting individual domestic horses grazing in the middle of a band of Takhis, Wit had to alert park staff to have the horses removed.

This is only one of many obstacles that Hustai managers encountered.

The first, surprisingly, was the task of figuring out how to acclimate the newly arrived Takhi into their Mongolian habitat, which was much larger than the pastures they'd previously lived in, and which held many new dangers. Even with the preparation of having the horses live first in semi-reserves, the survival of the Hustai Takhi was at first touch and go.

Since the horses couldn't simply be taken to Hustai and released, staff devised yet another interim plan. After the horses were flown from Europe to Mongolia, they traveled by truck to Hustai and were released in a fenced-in area where people could still watch them closely. At first the horses didn't know what to eat, so staff supplied food. The horses also needed to learn what predators lived in their new territory and how to defend themselves from those predators. One newly arrived group, still in the fenced area, had to learn this quickly. Soon after their arrival, a wolf jumped the fence and attacked them. The stallion and others drove the wolf off, and there were no disasters that night. When I visited Hustai, an older staff member who had been there that night told me this story, his face glowing with pride.

This strategy of slow release worked well, but then Hustai staff encountered yet another obstacle. When the gates in the fences were opened, staff expected the horses to leave their enclosure. However, the horses refused to go. They preferred to stay where they felt most secure. Herdsmen tried to tempt the mares by putting food just outside the gate, but the band stallion would not allow the mares to leave. Each time a mare approached the open gate and the food, the stallion drove her back inside. Nor could herdsmen get the stallion to leave. A rider on a domestic mount entered the enclosure and tried to entice the stallion to give chase. He did—up to a point. When the horse and rider galloped through the gate, the stallion stopped. Horses, even the Takhi, have comfort zones.

Park staff were at a loss. They had expected the Takhi to embrace freedom. Finally, staff hit on a solution: they removed the fencing. This way, the horses could stay where they were and explore the park at their own pace. In the early days of this phase of reintroduction, the horses didn't venture far. But eventually they became used to the area and began moving farther and farther away, lured, just as Whisper had been, by the promise of better grass.

Now, after twenty years, the horses have regained their survival skills and behave much like the horses of the Pryor Mountains and McCullough Peaks. The stallions fight with each other over the mares, and the mares sometimes object to the stallions' presumptions of power. In one case, after a young stallion won a battle with an older stallion, the usurper tried to control a group of mares by chasing a foal. The mares were having none of it. Herdsmen watched while the mares joined together and drove the upstart stallion off. Herdsmen also saw two young stallions, bachelor band buddies, team up to drive off an older stallion. The younger of the pair harassed the old stallion while the older of the pair took the mares. No fight ensued.

Keeping the domestic horses separate from the Takhis is an ongoing problem, not just because of local herdsmen, but also because of the Takhis themselves. At the tourist camp, a string of domestic horses is kept for tourists and staff to ride. One older Takhi stallion, driven off from the other Takhi, made a habit of coming into the tourist camp at night to round up the domestic mares and run them off into the mountains.

"Just to get himself some company," according to the park director.

More Takhi have been brought to Hustai since the first release, and many of those have been able to successfully raise foals. For the most part, the horses are left alone, although herdsmen know where they are. The horses are not typically fed, but if there are special circumstances, like a particularly severe winter, staff will supply fodder.

Hustai's rewilding project has been so successful that it's become a model, not just of how to return horses to the wild, but of how to return many kinds of animals to the wilderness. At the conference in Vienna where I met Jason Ransom and Laura Lagos, staff from Hustai presented information on the increase in survival rates of their horses over time. They received a standing ovation from the other conference attendees.

The winter of 2009–2010 was particularly severe in central Asia. There had been a drought before winter set in, so forage was already limited. Following that summer drought, heavy snows fell. Whatever food was available was difficult for the Takhi to get to. In the Gobi Desert, where the ITG had placed the second population of Takhi, the population of 138 was reduced to only 49 by winter's end. In a talk during the conference, scientists explained that mortality rates for the Gobi horses showed clear patterns. Zoo-raised Takhi experienced greater mortality than Takhi born in the desert region. The disastrous winter weather had another important long-term effect: only one mare delivered a foal the following spring. In a scientific paper about the Gobi disaster entitled "The Danger of Putting All Your Eggs in One Basket," the authors warned that another bad winter could completely wipe out the whole Gobi population.

By contrast, Hustai experienced fewer mortalities. By the time of that terrible winter, Hustai had 259 horses, many of whom had been born in the wild. Only about 10 percent of the horses died. There was a reduction in the number of foals born that spring, but overall survival rates were good.

I asked Piet Wit why he thought Hustai's survival rates had been so impressive. He suggested that the success might have been due in part to the varied terrain in the park. There were many refugia available to the Hustai horses, whereas the Gobi horses were trying to survive in a flat, open, and somewhat uniform land that offered the horses few options. The Gobi is a much harder place for horses to survive, since, as in the Australian outback, they have to travel much longer distances to find food and water. By contrast, Hustai has mountains, valleys, some forest, and some open land, as well as a large river that flows year-round. Just like in Spain during the Pleistocene, the Hustai horses had choices.

I also asked Bandi Namkhai, the park director, about why *he* thought the park had been so successful during that winter. He attributed it to the skills of local herders, who have been looking after livestock for generations. Caring for animals, Bandi suggested, is just in their blood. Herders responsible for the well-being of the Hustai Takhi keep a close watch on their charges. The horses are "wild" in a sense, but the rangers know where each horse is and what that horse is up to. In contrast, the Gobi Takhi range over a wider area and are not tracked on a daily basis.

After the conference, Piet Wit and I decided to get a full view of the area by ascending the sacred peak of Hustai Mountain, near the center of the park.

Mongolia is a place where the highest mountains are home to the most powerful gods, and as we lumbered up the rocky slopes in a spine-jolting Russian-manufactured four-wheel-drive transport truck, I decided that this particular mountain must have been home to a genuinely mighty deity. The rocky ascent went on and on.

Just before we got to the peak, we dismounted and walked, leaving our rattling gray machine to rest where, for thousands of years, Mongolians have left their horses to graze. As we walked, we saw around us a majestic world bounded by ranges of purple mountains off in the distance. We watched strange ducks nesting in the high mountains far from any water. We saw scores of red deer on high alert, monitoring our travel. Above us, steppe eagles soared, searching for prey. Somewhere out there far below us were the Takhi.

At the top, Piet and I circled three times around a sacred Buddhist *ovoo*, a rock cairn bedecked with silk prayer flags of many different colors, as well as with shoes, crutches, paper money, and anything else that people have felt like leaving here.

Standing with Piet, I looked south toward the fabled Gobi Desert and felt at my back a hint of Nettie's easy-chair wind. This time the wind flowed down not from Canada but from Siberia, but it was all part of the same chilly Arctic Circle weather system that dominates our Northern Hemisphere and that would bring, later that day, just as it had in Wyoming, a pelting hailstorm that would cause us, once again, to break for shelter.

Up on the mountain that day, though, breathing in the thin, clean air, I looked up at the bluest sky I'd ever seen. I saw below me a world free of fences, of paved roads, of tall buildings, or of any buildings at all, for that matter. According to Mongolian law, most of this land will never be privately owned. It belongs to the horses and to the people and is intended to stay forever open.

I saw a mesmeric world of rolling hills that looked quite like Wyoming might once have looked. There were arid steppe grasslands dotted with stands of trees. Brush grew in the damper ravines, and wildlife trails twisted around the sides of the steep slopes. Hustai's Tuul River valley far below us still contains precious four-thousand-year-old graves holding the remains of some of the world's earliest horsemen—men who galloped their horses over the open plains of Asia long before the fabled rule of Genghis Khan. Change is coming to Mongolia in the form of mining and international commerce, and the horses and people will have much to contend with. But I wasn't worried about that then, as I stood with Piet.

After all, in Mongolia, in the summertime, when the grass is growing and the rivers are full and the sun is high—what else would you choose to be but a nomad? Chris Norris had been right to tell me: "Horses have a story to tell." The story is one of resilience and flexibility in response to an ever-changing planet, of sociability and intelligence, and of partnership with the human race. It's a thrilling story, one with many more layers than those peeled back by Charles Darwin, and one that no doubt has many more layers yet to be understood.

I only wish, for the sake of his health, that poor old Charles Darwin had had a crystal ball. He would have been a much happier man.

EPILOGUE: BACKYARD MUSTANG

Phyllis Preator and I stood on a ridge in the McCullough Peaks horse management area, looking down into a multihued canyon that stretched as far as our eyes could see, where time was exposed with almost unfathomable complexity. In her younger days Phyllis spent a lot of time in that endless space, sometimes searching for horses, sometimes just riding, and it still thrills her. Places with too many trees make her feel closed in.

"This place is as grand as the Grand Canyon," I said. From my East Coast point of view, there wasn't much grass for grazing, but there was certainly a lot of room to roam.

"Do you think so? We do." She was clearly pleased.

We stood in the hunnert-degree heat for a while. For once, Nettie Kelley's easy-chair wind had died down and there didn't appear to be a hailstorm on the horizon. At least, not yet.

"Phyllis," I said, "do you think a Wyoming mustang could find happiness in a New England pasture?"

Her face lit up.

"Would you pay a lot of attention to him? Would you spend every day with him and find him other horses to be with and really take care of him?"

"Yes," I said. "I would."

"Then," she said, "I think he'd be very happy there."

So I filled out the forms.

AUTHOR'S NOTE

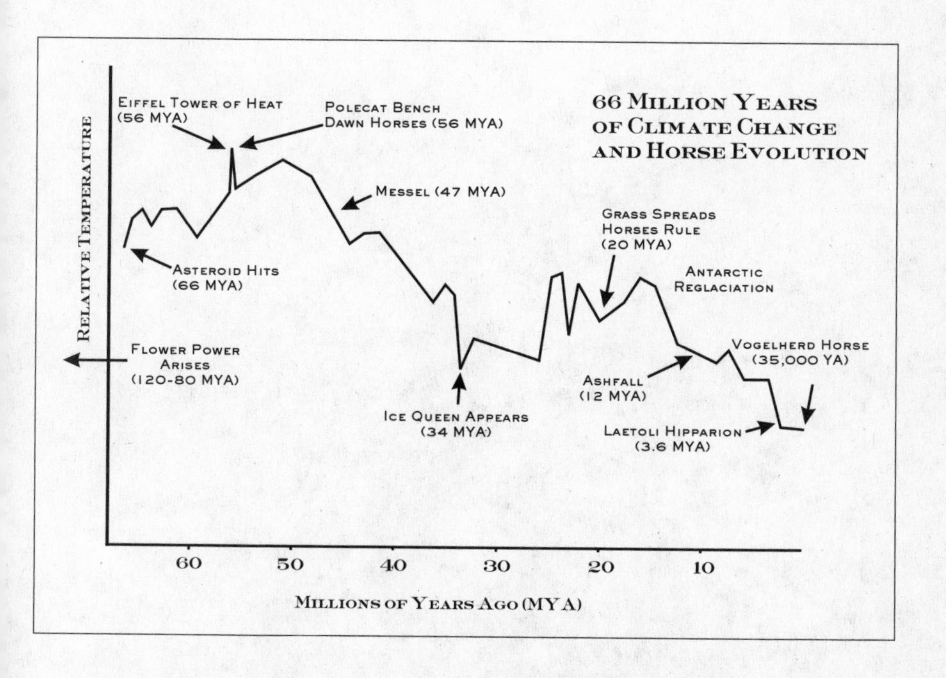

How is it that Whisper became my charming companion for a short while when I was young? As I researched this book over three years, the intimate relationship between the planet's major energy systems, like tectonics and ocean currents, became increasingly clear. Then I realized that these systems created climate shifts that profoundly altered life on the planet, in terms of both plant life and animal life. Particularly in herbivores (like the horse), evolution responded, sometimes slowly and sometimes sharply, to atmospheric composition and movement, to heat and cold, to rainfall patterns, and to new forms of plant life.

I drew a rough draft of 66 million years of climate change (based on more scientific drawings) and placed many of the major evolutionary events discussed in this book in that context, showing how major climate events coincided with important events in the evolution of horses, humans, plant life, and ocean currents. I thought of it as an Easter egg hunt. Over the past decade or so, scientists in many different fields have ferreted out all this data, and it was immensely satisfying to see how it all fit together. I haven't had so much fun since I galloped Whisper over Vermont's back roads many, many decades ago.

NOTES

1. Watching Wild Horses

11 Wild Horses: Which horses in the world are truly wild, and which only "feral"? I began writing about "feral" horses—horses that supposedly come from domesticated stock that, some time in the historical past (as opposed to the prehistoric past), escaped human domination and began living on their own. In the view of some researchers, there are no truly "wild" horses left in the world, save for a small group of horses called the Przewalski's or Takhi horse in Mongolia, which I discuss in greater depth in chapter 10. However, the more I learned about horses, the more certain I became that, in the case of many of the world's bands of free-roaming horses, the details of their ancient history remain a mystery. No one knows how, or whether, domestication changed the basic nature of the horse. After I read a paper by Jonaki Bhattacharyya et al., "The 'Wild' or 'Feral' Distraction: Effects of Cultural Understandings on Management Controversy over Free-Ranging Horses (*Equus ferus caballus*)," *Human Ecology* 39, (2011), http://link.springer.com/article/10.1007%2Fs10745-011-9416-9, I decided to forgo the distinction and simply use the term "wild" to describe free-roaming horses worldwide. The differentiation between "feral" horses and "wild" horses implies a rigid boundary that simply doesn't exist in many of the world's cultures.

11 "There is no doubt": William Beebe was one of the world's most popular naturalists in the twentieth century. The author of a myriad of books, he was one of the first to dramatize science for the general public when he narrated his descent into the depths of the ocean in a bathysphere for a radio audience. This quote was recounted to me by the paleontologist Eric Scott of the San Bernardino County Museum.

11 Sometime around thirty-five thousand years ago: We who are used to the certainty of the historical age are bound to be constantly frustrated by the vague dates of prehistory. They're just something we're going to have to learn to live with. The age of the Vogelherd horse is hotly debated. Some researchers suggest that

it may be only thirty thousand years old, while others suggest it may be older than thirty-five thousand years. This is the case with nearly all dates referring to events in prehistory. As a consequence, I have taken what I believe to be the majority consensus and then rounded off those dates.

11 "an abstraction of the graceful essence of the horse": Ian Tattersall, *Masters of the Planet: The Search for Our Human Origins* (New York: Palgrave Macmillan, 2012), 180.

12 "esthetically perfect": Harald Floss, phone conversation with author, April 22, 2014.

14 the lines are now well-worn: Harald Floss, phone conversation with author, April 22, 2014.

14 the most frequently represented animal: I first read this in Christine Desdemaines-Hugon's *Stepping-Stones: A Journey through the Ice Age Caves of the Dordogne* (New Haven: Yale University Press, 2010). I admit to having been skeptical, but numerous scholarly tomes have confirmed the assertion.

15 On the walls of Chauvet Cave: Chauvet Cave is not open to the general public, but numerous books, videos, and photos on the Web are available to those who are interested.

20 a region known as: As a result of the Wild Free-Roaming Horses and Burros Act of 1971, the federal government protects wild horses as "living symbols of the historic and pioneer spirit of the West" on several hundred designated properties across the United States. Some of the horses are under the protection of the U.S. Bureau of Land Management, of the U.S. National Park Service, or of the U.S. Forest Service. For more information on the bitter politics that accompany these horses wherever they roam, see Hope Ryden's *America's Last Wild Horses* (Guilford, CT: Lyons Press, 2005). I've chosen not to delve into this complicated political subject but rather to focus on the science of the horse-human partnership.

24 mares huddling together: See H. M. Peel's *Fury, Son of the Wilds* (Franklin Watts, 1959) for a classic example of this type of story. Peel relates the story of an Australian brumby stallion in these antiquated terms in great detail, but this kind of thing can also be found in numerous books published in the twentieth century for children that tell stories of individual horses living wild. In these tales, the mares are almost always helpless and the band stallions generally fairly noble and protective. I suspect the misunderstanding stems from the glamour that stallions so frequently display. While decidedly less glamorous, the mares, it seems, are the true engine of the band's cohesiveness.

26 "a dominant stallion, subordinate adult males and females": National Research Council of the National Academies, *Using Science to Improve the BLM Wild Horse and Burro Program: A Way Forward* (Washington, D.C.: The National Academies Press, 2013), 27. www.nap.edu/catalog.php?Record_id=13511.

26 The ecologist Joel Berger studied: Joel Berger, *Wild Horses of the Great Basin: Social Competition and Population Size* (Chicago: University of Chicago Press, 1986).

28 "neither the dominant nor the most aggressive animals": Katherine A. Houpt,

Ronald Keiper, "The Position of the Stallion in the Equine Dominance Hierarchy of Feral and Domestic Ponies," *Journal of Animal Science* 54 (1982): 945–50.

28 the British researcher Deborah Goodwin: Deborah Goodwin, "The Importance of Ethology in Understanding the Behaviour of the Horse," *Equine Veterinary Journal* 28 (1999): 15–19.

35 more than a million free-ranging horses in the Australian outback alone: These figures are hotly debated, particularly since the government-authorized culling of the horses. As with most wildlife populations, no one really knows how many of these horses, called "brumbies," roam the island continent, but it is obvious that there are a lot. Horse-vehicle collisions are common.

38 white horses have an advantage: Gábor Horváth et al., "An Unexpected Advantage of Whiteness in Horses: The Most Horsefly-Proof Horse Has a Depolarizing Coat," *Proceedings of the Royal Society B* 277 (2010): 1643–50.

38 Canada's Sable Island: Canadian researcher Philip D. McLoughlin spoke about the Sable Island horses and evolutionary theory at the conference discussed in the text. Following his presentation, Dr. McLoughlin was kind enough to discuss his research with me in depth both at the conference and several times on the telephone. He also forwarded me a number of his research studies as well as authoritative material on the background of the horses. Most of the information on Sable Island comes from those discussions and printed research. However, the theory about the shortening of the horses' pasterns in response to the need to climb the steep sand dunes is mine alone.

40 The Sable Island horses are also behaviorally unusual: Adrienne L. Contasti et al., "Explaining Spatial Heterogeneity in Population Dynamics and Genetics from Spatial Variation in Resources for a Large Herbivore," *PLoS One* 7 (2012), www.plosone.org/article/info%3Adoi%2F10.1371%2Fjournal.pone.0047858.

41 in the Yana River region: Pitulko, V. et al., "The Yana RHS Site: Humans in the Arctic before the Last Glacial Maximum," *Science* 303 (2004): 52–56.

41 "the best running animal on the planet": Darrin Pagnac, personal communication.

44 "As long as the wild horses continue to roam": J. Sanford Rikoon, "Wild Horses and the Political Ecology of Nature Restoration in the Missouri Ozarks," *Geoforum* 37 (2006): 200–11.

2. In the Land of Butch Cassidy

45 "If you want to sense the evolution of the modern horse": Richard Tedford of the American Museum of Natural History was one of a group of scientists who gathered at the museum in 1981 to discuss the phenomenal spread of *Hipparion* horses. An article about the conference, titled "The Hipparion Is Still an Elusive Horse," by Bayard Webster, ran in *The New York Times* on November 17, 1981. It is accessible here: www.nytimes.com/1981/11/17/science/the-hipparion-is-stilll-an-elusive-horse.html.

45 seas covered what is now Wyoming: One of the best ways to understand how the unique topography of the American West came about is to spend a day at

the Denver Museum of Nature and Science. This museum has all kinds of exhibits, including a presentation of the geological forces that over hundreds of millions of years shaped the landscape that we experience today.

45 continental drift, tectonic collision: For a great, accessible discussion of plate tectonics, read Simon Winchester's *A Crack in the Edge of the World: America and the Great California Earthquake of 1906* (New York: HarperCollins, 2005).

45 "shale so black it all but smelled of low tide": John McPhee, *Rising from the Plains* (New York: Farrar, Straus and Giroux, 1986), 11.

47 The result is her book: Phyllis Preator, *Facts and Legends: Behind the McCulloch Peaks Mustangs* (self-published, 2012). ISBN 978-0-692-01509-4.

50 the modern horse: We can see vestiges of the common ancestry of humans and horses in our skeletons. We both have tarsals and metatarsals, fibulas and tibias, and even patellae—all evidence of our biological kinship. We also share the calcaneus bone, though the bone takes on a different form in horses than in humans. In humans it is the heel bone, but in the horse it is located in the hock joint, halfway up the horse's hind leg.

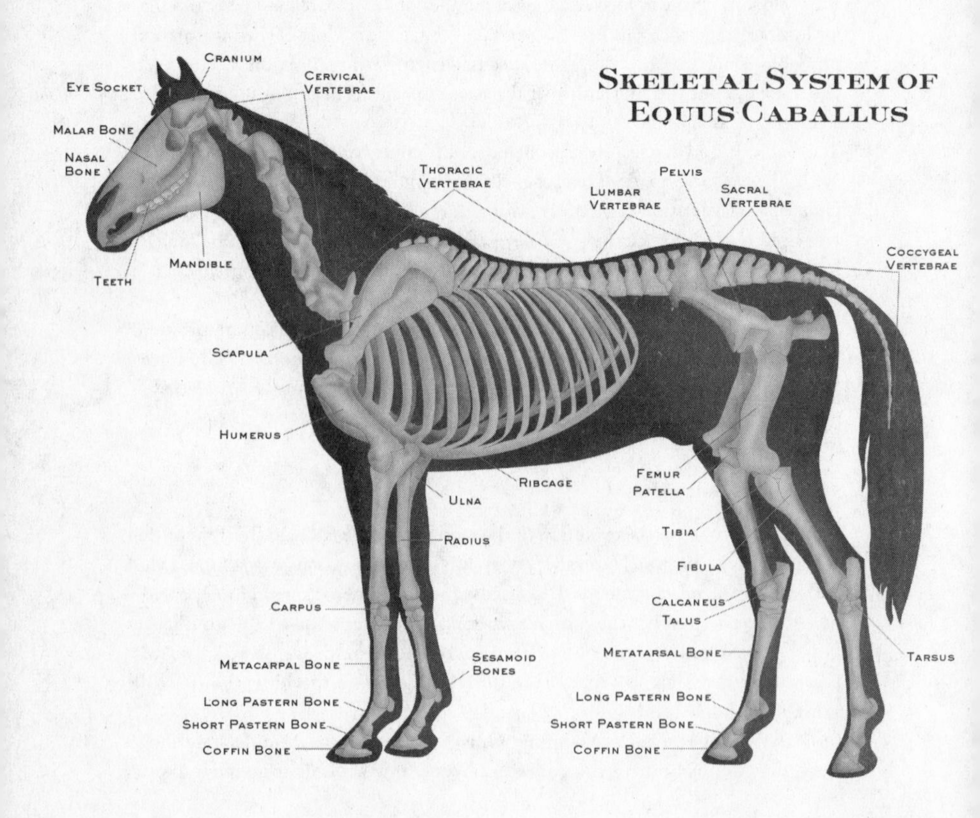

52 "artful dodgers": Christine Janis, "Victors by Default: The Mammalian Succession," in *The Book of Life: An Illustrated History of the Evolution of Life on Earth*, 2nd ed., ed. Stephen Jay Gould (New York: W. W. Norton Company, 2001), 171.

52 *Repenomamus*: Yaoming Hu et al., "Large Mesozoic Mammals Fed on Young Dinosaurs," *Nature* 433 (2005): 149–52.

53 A 2010 paper in *Science*: Peter Schulte et al., "The Chicxulub Asteroid Impact and Mass Extinction at the Cretaceous-Paleogene Boundary," *Science* 327 (2010): 1214–18.

53 "asteroid porn": Chris Norris, "With Neither Bang, Nor a Whimper," blog entry, *Prerogative of Harlots*, April 28, 2009, http://paleocoll.blogspot.com/2009/04/with-neither-bang-nor-whimper.html.

53 "Don't get me wrong": This statement derives from a conversation I had with Archibald, but an in-depth discussion of the early days of mammalian expansion and the paleontological point of view regarding the effect of the Chicxulub asteroid can be found in David Archibald's book *Extinction and Radiation: How the Fall of Dinosaurs Led to the Rise of Mammals* (Baltimore: Johns Hopkins University Press, 2011).

54 "victors by default": Janis, "Victors by Default," 173.

55 For a brief period, it was *very* hot: Evidence for the Paleocene-Eocene Thermal Maximum is by now multitudinous. For a good basic description, read Phil Jardine's "Patterns in Palaeontology: The Paleocene-Eocene Thermal Maximum," *Paleontology Online* 1 (2011): 5, www.palaeontologyonline.com/articles/2011/the-paleocene-eocene-thermal-maximum/.

56 horses may well have originated right there: I first read this in Robert Kunzig's article "World Without Ice," *National Geographic*, October 2011. It's an unusual idea, so just to be sure, I spoke to Gingerich himself and asked if he still believed this. "Why not?" he answered.

61 the paleontologist Ken Rose compared: Much of the background material on both these early horses and these early primates comes from Kenneth D. Rose's comprehensive textbook, *The Beginning of the Age of Mammals* (Baltimore: Johns Hopkins University Press, 2006).

62 gave Charles Darwin a serious headache: There are, of course, numerous biographies of Darwin. My favorite, widely believed to rank among the most definitive, is Adrian Desmond and James Moore's immense *Darwin: The Life of a Tormented Evolutionist* (New York: Warner Books, 1992). For the discussion of Darwin's health during certain times in his life, as well as his proclivity for visiting health spas, I have relied on this text.

64 experienced an earthquake: Marcia Bjornerud, *Reading the Rocks: The Autobiography of the Earth* (New York: Westview Press, 2005).

64 when the English genius Thomas Henry Huxley: This story has been told in hundreds of books and articles. It was also related to me directly by Chris Norris when I visited the Yale Peabody Museum.

68 about the size of a small dog: Ross Secord et al., "Evolution of the Earliest Horses Driven by Climate Change in the Paleocene-Eocene Thermal Maximum," *Science* 335 (2012): 959–62.

68 horses made changes: Philip D. Gingerich, "Variation, Sexual Dimorphism, and Social Structure in the Early Eocene Horse *Hyracotherium* (Mammalia, Perissodactyla)," *Paleobiology* 7 (1981): 443–55, www.jstor.org/stable/i317541.

3. The Garden of Eden Appears, Then Vanishes

73 "Living horses, with their high-crowned grinding teeth": Michael Novacek's *Dinosaurs of the Flaming Cliffs* (New York: Doubleday, 1996) is a terrific adventure story that makes for fun reading, even for those who are not deeply interested in paleontology. Accessible and easy to understand, the book is more about the paleontological lifestyle than about dinosaurs themselves. Mike Novacek is not only a respected researcher, but a good writer.

74 Grube Messel: Anyone even remotely interested in geology, evolution, or horses would benefit greatly by a trip to Messel. There is a museum on the site, and during the warmer months guided tours are available. Unfortunately, very little information is available to visitors who do not speak German. A good introductory English-language book is Jens Lorenz Franzen's *The Rise of Horses: 55 Million Years of Evolution*, translated by Kirsten M. Brown (Baltimore: Johns Hopkins University Press, 2010). This book has wonderful photos of some of the horse fossils found in the pit.

77 nicknamed Ida: A great deal of controversy surrounds the fossil dubbed Ida, regarding exactly where on the evolutionary bush of life the fossil belongs—whether Ida is our direct ancestor or had already taken a baby step or two down a different twig. To understand the thrill of discovering Ida, check out Colin Tudge and Josh Young, *The Link: Uncovering Our Earliest Ancestor* (New York: Little, Brown and Company, 2009).

81 the Greek tale of Orpheus and Eurydice: Eurydice, Orpheus's wife, died from a snake bite immediately after her wedding. Orpheus, a real charmer, convinced the gods to allow him to go into Hades and bring her back to the land of the living. Unfortunately, as soon as he emerged from Hades, he turned to look at Eurydice, who vanished just as soon as he saw her.

85 this horse, *Orohippus*, still had four toes on his front feet: For a great, and authoritative, synopsis of early horse paleontology in Wyoming, including the discovery of *Orohippus* at Grizzly Buttes, check out the American Museum of Natural History's extensive online information, which has both old photos of the expeditions and written text: http://research.amnh.org/paleontology/photographs/1905-wyoming-eocene/.

88 I had read about *Epihippus* once before: Tim Flannery, *The Eternal Frontier: An Ecological History of North America and Its Peoples* (New York: Grove Press, 2002), 100.

93 horses changed their diet: Matthew C. Mihlbachler et al., "Dietary Change and Evolution of Horses in North America," *Science* 331 (2011): 1178–81.

94 "Radinsky's brains": Leonard Radinsky, "Oldest Horse Brains: More Advanced Than Previously Realized," *Science* 194 (1976): 626–27.

95 "Evolution of the Horse Brain": Tilly Edinger, "Evolution of the Horse Brain," *The Geological Society of America Memoirs* 25 (1948): 1–177. For an excellent and concise overview of Edinger's remarkable life and effect on paleontology, read Emily A. Buchholtz and Ernst-August Seyfarth, "The Study of 'Fossil Brains': Tilly Edinger (1897–1967) and the Beginnings of Paleoneurology," *BioScience* 51 (2001). Available in full online: http://bioscience.oxfordjournals.org/content/51/8/674.full.

96 "considerably advanced": Radinsky, "Oldest Horse Brains."

4. The Triumph of *Hipparion*

99 "The history of the horse family": George Gaylord Simpson's book *Horses: The Story of the Horse Family in the Modern World and Through Sixty Million Years of History*, published in 1951, was a popular international bestseller that thoroughly explained the evolution of horses. Since then, science has of course progressed considerably, so that much of what Simpson explained is now more thoroughly understood, particularly how the evolution of horses has been driven by tectonics and climate.

99 about 3.6 million years ago: The information in this narrative section is derived from numerous studies of the Laetoli ecosystem of that time frame. There has been extensive documentation by a variety of researchers who have looked not only at the early horses and humans and all other animal footprints present at the site, but also at the makeup of the plants in the area and at the geology and climate. My narrative also relies on discussions with several scientists, including Andrew Hill of Yale University, who found the tracks, and with Elise Renders, retired researcher and avid equestrian, whose fabulous analysis of the tracks made by the Laetoli mare and foal revealed for the first time how *Hipparion*'s "extra" toes helped stabilize the horse's gait, as well as the fact that, in at least some horses, the four-beat gait is fully natural: Elise Renders, "The Gait of *Hipparion* sp. from Fossil Footprints in Laetoli, Tanzania," *Nature* 308 (1984): 179–81.

100 the clueless foal ran right in front of the mare: Remarkably, the behavior of this young animal was recorded in the ash and can still be seen today by closely studying the tracks.

100 a small band of our own ancestral relatives: The discovery of these footprints has been heralded worldwide. When I visited the Musée National de Préhistoire (the National Museum of Prehistory, well worth devoting at least a half day to) in Les Eyzies, France, I found that at the entrance to the exhibits lay a replica of these preserved footprints. Research analyzing these footprints is abundant, but it's fun to look at the original publication announcing the find, now available for all to see: M. D. Leakey and R. L. Hay, "Pliocene Footprints in the Laetolil Beds at Laetoli, Northern Tanzania," *Nature* 278 (1979), www.nature.com/nature/ancestor/pdf/278317.pdf.

100 *Australopithecus afarensis*: For a clear, succinct discussion—highly accessible to the lay reader—of both the importance of this ancestral relative of ours and the

influence of the discoveries made at Laetoli, see Martin Meredith's *Born in Africa: The Quest for the Origins of Human Life* (New York: PublicAffairs / Perseus, 2011).

103 a two-volume set of papers: Mary D. Leakey and J. M. Harris, eds., *Laetoli: A Pliocene Site in Northern Tanzania*, Oxford Science Publications (Oxford, U.K.: Clarendon Press, 1987).

107 quickly multiplied into countless species: W. A. Berggren and J. A. van Couvering, "The Late Neogene, Vol. 2: Biostratigraphy, Geochronology and Paleoclimatology of the Last 15 Million Years in Marine and Continental Sequences," *Developments in Palaeontology and Stratigraphy* (Amsterdam: Elsevier Scientific Publishing, 1974).

107 excelled at evolutionary experimentation: Countless species of these horses thrived all over the world, except for Antarctica, for more than 20 million years. To sort them all out has been an impossible task for paleontologists, but several people have tried. A heroic attempt, now rather dated, is Bruce J. McFadden's *Fossil Horses: Systematics, Paleobiology, and Evolution of the Family Equidae* (Cambridge, U.K.: Cambridge University Press, 1992).

107 "one of the great animal travelers": Simpson, *Horses*, 140.

108 studied leaf waxes of plants growing in Antarctica: Sarah Feakins oversees her own Leaf Wax Lab at the University of Southern California, where she specializes in studying the paleontological history of leaf waxes from all over the world. The paper discussed here is Sarah J. Feakins et al., "Hydrologic Cycling over Antarctica During the Middle Miocene Warming," *Nature Geoscience* 5 (2012): 557–60, www.nature.com/ngeo/journal/v5/n8/full/ngeo1498.html.

109 "higgledy-piggledy": John Herschel, astronomer, mentioned in a letter from Charles Darwin to Charles Lyell, dated December 10, 1859. Viewable online at Darwin Correspondence Project, www.darwinproject.ac.uk/letter/entry-2575.

110 the paleontologist Gregory Retallak: Gregory J. Retallack, "Cenozoic Expansion of Grasslands and Climatic Cooling," *The Journal of Geology* 109 (2001): 407–26, www.jstor.org/discover/10.1086/320791?uid=3739696&uid=2&uid=4&uid=3739256&sid=21104007857767.

113 both kinds of grasses ebbed and flowed over the landscape: Sarah J. Feakins et al., "Northeast African Vegetation Change over 12 M.Y.," *Geology*, published online January 17, 2013, http://geology.gsapubs.org/content/early/2013/01/17/G33845.1.abstract.

113 some African animals changed in response to the new grasslands: Kevin T. Uno et al., "Late Miocene to Pliocene Carbon Isotope Record of Differential Diet Change Among East African Herbivores," *Proceedings of the National Academy of Sciences* 108 (2011): 6509–14, DOI:10.1073/pnas.1018435108.

120 the triumph of one-toed horses: Nicholas A. Famoso and Darrin Pagnac, "A Comparison of the Clarendonian Equid Assemblages from the Mission Pit, South Dakota and Ashfall Fossil Beds, Nebraska," *Transactions of the Nebraska Academy of Sciences and Affiliated Societies* 32 (2011): 98–107, http://digitalcommons.unl.edu/cgi/viewcontent.cgi?article=1008&context=tnas.

120 Idaho's Hagerman Horse Quarry: Dean R. Richmond and H. Gregory McDonald, "The Hagerman Horse Quarry: Death and Deposition," written for the National Park Service and published online at www.nature.nps.gov/geology/paleontology/pub/grd3_3/hag1.htm.

5. *Equus*

122 "A hoof is like a second heart in a horse": J. Edward Chamberlin, *Horse: How the Horse Has Shaped Civilizations* (Katonah, NY: Blue Bridge, 2006), 73. This book is filled with heart-stoppingly beautiful sentences.

122 The last meal of the golden-coated Yukon horse was buttercups: This narrative section is based on extensive reading as well as on a number of conversations with Grant Zazula of the Yukon Beringia Interpretive Centre, operated by the Yukon government. The centre is both a museum and interpretive center for the public and a gathering spot for scientists interested in research in the local area. Information about the research is available online at www.beringia.com/index.html.

122 the Yukon horse: Grant Zazula and Duane Froese, *Ice Age Klondike: Fossil Treasures from the Frozen Ground* (Whitehorse: Government of Yukon, 2011). This booklet, written by respected scientists, is an excellent introduction to the world of the Yukon horse. It's available online at www.tc.gov.yk.ca/publications/ice_age_klondike_2011.pdf.

126 the "Big Three" mammals: R. Dale Guthrie, "Mammals of the Mammoth Steppe as Paleoenvironmental Indicators," in *Paleoecology of Beringia*, ed. David M. Hopkins, John V. Matthews, Charles E. Schweger, and Stephen B. Young (New York: Academic Press, 1982).

130 "a flickering switch": Stephen Barker, "The 'Flickering Switch' of Late Pleistocene Climate Change Revisited," *Geophysical Research Letters* 32 (2005): C1583, http://onlinelibrary.wiley.com/doi/10.1029/2005GL024486/pdf.

131 ongoing chaos in the interior of the American West: Donald Grayson, *The Great Basin: A Natural Prehistory*, rev. and exp. ed. (Berkeley and Los Angeles: University of California Press, 2011).

133 mammoths, and mastodons: For a great discussion of the impact that mastodons had on Thomas Jefferson and on the early years of American science, read Stanley Hedeen, *Big Bone Lick: The Cradle of American Paleontology* (Lexington: University Press of Kentucky, 2008).

133 "man and man alone": Paul S. Martin, "Prehistoric Overkill," in *Pleistocene Extinctions: The Search for a Cause*," Paul S. Martin and H. E. Wright, Jr., eds. (New Haven: Yale University Press, 1967).

135 the archaeologist Dennis Jenkins has found: Dennis L. Jenkins et al., "Clovis Age Western Stemmed Projectile Points and Human Coprolites at the Paisley Caves," *Science* 337 (2012): 223–28, DOI: 10.1126/science.1218443.

136 Wally's Beach: Brian Kooyman et al., "Late Pleistocene Horse Hunting at the Wally's Beach Site (DhPg-8), Canada," *American Antiquity* 71 (2006): 101–21.

137 Gary Haynes believes: Gary Haynes, "Extinctions in North America's Late Glacial Landscapes," *Quarternary International* 285 (2013): 89–98, www.unr.edu/Documents/liberal-arts/anthropology/gary-haynes/Haynes_QI_Extinctions2012Reprint.pdf.

137 Haynes has even taken a stab: Gary Haynes of the University of Nevada gave a lecture on the disappearance of the horses for the general public at the Royal Tyrrell Museum in Alberta, Canada, in February 2012. It's available on YouTube at www.youtube.com/watch?v=8WZ5Q2JYbLY.

138 "Size matters": For a short, delightful, and accessible read on this subject, see John Tyler Bonner, *Why Size Matters: From Bacteria to Blue Whales* (Princeton: Princeton University Press, 2006). And for further information, see Bonner's *Randomness in Evolution* (Princeton, 2013), where he reiterates the role that random chance plays in evolution.

6. The Arch of the Neck

143 "The steady thunder": Tamsin Pickeral, *The Horse: 30,000 Years of the Horse in Art* (London and New York: Merrell, 2006), 20. Pickeral provides a fascinating look at the evolution over time of horses in human art.

143 Spain began attacking France: For an excellent and poetic explanation of how our planet's numerous plates collide, read Winchester, *A Crack at the Edge of the World* 149–56.

143 Iberia is said to have been a refugium: There are numerous papers that discuss this, but the trend of late is to look more closely at specific sections of Iberia. For a start, try Gonzalo Nieto Feliner, "Southern European Glacial Refugia: A Tale of Tales," *Taxon* 60 (2011): 365–72, http://digital.csic.es/bitstream/10261/35607/1/2011_Nieto_Taxon60%282%29.pdf.

146 many of these mouths were occupied: Lawrence Guy Straus, *Iberia Before the Iberians: The Stone Age Prehistory of Cantabrian Spain* (Albuquerque: University of New Mexico Press, 1992; repr. ed., 2011).

150 Neanderthals and the first *Homo sapiens*: Recent dating of several Neanderthal sites has challenged the long-accepted belief that Neanderthals lived longer in Iberia than elsewhere, but the battle isn't over yet. For some background, see Ewen Callaway, "Neanderthal Settlements Point to Earlier Extinction: New Dating Suggests Bones from Spanish Sites Are 10,000 Years Older Than Previously Thought," *Nature* News, February 4, 2013, www.nature.com/news/neanderthal-settlements-point-to-earlier-extinction-1.12355.

150 hunted horses using different techniques: Straus, *Iberia Before the Iberians*.

159 "graffiti": For anyone interested in Pleistocene art, R. Dale Guthrie's book *The Nature of Paleolithic Art* (Chicago: University of Chicago Press, 2006) is a must-read. Guthrie's view is entirely outside the bounds of accepted dogma, which makes it a fascinating counterweight to that of many more-conventional thinkers.

160 "those who praise horses": Chamberlin. *Horse*, 47.

160 very primitive stones: This finding is very controversial. Several research teams

have engaged in pitched battles over the issue, showing how difficult it is to analyze human behavior in the deep past. I once walked over a very dry area of Zimbabwe with a local expert in early stonework. "The ground is littered with tools," he explained, picking up rock after rock. But where he saw tools, I saw nothing more than sharp rocks. For a good overview of the argument, see Kate Wong, "Did Lucy's Species Butcher Animals?", *Observations*, *Scientific American* blog, April 13, 2011, http://blogs.scientificamerican.com/observations/2011/04/13/did-lucys-species-butcher-animals/.

160 clearly shows tools fashioned by *Homo erectus*: Jean de Heinzelin et al., "Environment and Behavior of 2.5-Million-Year-Old Bouri Hominids," *Science* 284 (1999): 625–29, www.indiana.edu/~origins/teach/P314/Bouri2.pdf. By now a plethora of papers have been written about this site, and most researchers agree that tools are present.

160 "You can almost imagine the stone waste materials": For an evocative description of this, read T. Douglas Price, *Europe Before Rome*: *A Site-by-Site Tour of the Stone, Bronze, and Iron Ages* (New York: Oxford University Press, 2013). This is another absolute must-read for people looking for an overview of early Europe and an excellent informal guidebook, since it is structured as a list of archaeological sites.

161 the oldest known hunting spears: Hartmut Thieme, "Lower Palaeolithic Hunting Spears from Germany," *Nature* 385 (1997): 807–10, www.nature.com/nature/journal/v385/n6619/abs/385807a0.html.

161 The archaeologist John Hoffecker theorizes: John F. Hoffecker, *Landscape of the Mind: Human Evolution and the Archaeology of Thought* (New York: Columbia University Press, 2011).

161 Hoffecker also studied a site on a floodplain of the Dnieper River: John Hoffecker, personal communication, and John F. Hoffecker et al., "Geoarchaeological and Bioarchaeological Studies at Mira, an Early Upper Paleolithic Site in the Lower Dnepr Valley, Ukraine," *Geoarchaeology* 29 (2014): 61–77, www.researchgate.net/publication/259538309_Geoarchaeological_and_Bioarchaeological_Studies_at_Mira_an_Early_Upper_Paleolithic_Site_in_the_Lower_Dnepr_Valley_Ukraine.

161 at another site, Kostenki: John F. Hoffecker et al., "Evidence for Kill-Butchery Events of Early Upper Paleolithic Age at Kostenki, Russia," *Journal of Archaeological Science* 37 (2010): 1073–89, www.researchgate.net/publication/229414951_Evidence_for_kill-butchery_events_of_early_Upper_Paleolithic_age_at_Kostenki_Russia.

7. The Partnership

164 "a man on a horse": John Steinbeck's 1937 novella *The Red Pony* is the best summation of the horse-human partnership that I know of.

164 died because of a climate anomaly in the warming Atlantic: This news made headlines around the world. The reported number of ponies who died varies

widely, and I've chosen one of the more conservative figures. For one such news report, see BBC News, "Acorn Glut Kills 90 New Forest Ponies and Cattle," January 15, 2014, www.bbc.com/news/uk-england-hampshire-25552107.

165 when the ice finally disappeared: To better understand what happened as the Pleistocene came to a close, I suggest reading a few of the books by the prolific Steven Mithen, including *After the Ice: A Global Human History* 20,000–5000 B.C. (Cambridge, MA: Harvard University Press, 2006). This book provides a good overall view of how the world got going again after the Pleistocene. Barry Cunliffe's *Europe Between the Oceans: 9000 B.C.–A.D. 1000* (New Haven: Yale University Press, 2008) is a fabulous book that will change the way you look at Europe.

165 Having originated long ago in what's now China: The story of how oak trees changed the world is compelling, deserving its own book-length treatment. During the heyday of the North American horses, there were oak trees, but their numbers were substantially limited. And yet, after the ice melted they were everywhere. For a brief discussion of how oaks spread over North America in particular, read Gene Stowe, "Research into Oaks Helps Us Understand Climate Change," Notre Dame News, July 23, 2012, http://news.nd.edu/news/32154-research-into-oaks-helps-us-understand-climate-change/.

167 covered by the encroaching sea: Vince Gaffney is a landscape archaeologist who has become well-known for his mapping of Doggerland, a region north of what we now think of as Europe's mainland. For a brief video overview of his work, go to www.birmingham.ac.uk/schools/historycultures/departments/caha/research/arch-research/videos/gaffney-vince-doggerland.aspx.

167 "hard-wired to be mobile": Cunliffe, *Europe Between the Oceans*, vii.

168 at a site called Formby Point: Local researcher Gordon Roberts, a member of the Sefton Coast Partnership Archaeology and History Task Group, a team of people who photograph these fleeting tracks as they are first revealed and then destroyed by the wave action of the ocean, replied in detail to my inquiry: "The horse hoof prints which I have recorded were not associated with any human footprints and, indeed, may well precede the arrival of hunter-gatherers on the coastline. Almost all of the so-called 'Formby Footprints' are located on the shore-most muddy margins of a long-vanished intertidal lagoon which lay between a series of sandy, barrier islands and the prehistoric coastline of 5000 B.C. to 3000 B.C. The horse hoof prints were preserved in an ancient, weather-hardened exposure of blue-grey marine mud, deposited—I surmise—about eight millennia ago, during an earlier period of the evolution of the coastline. There were no human footprints, nor—apart from a few red deer prints—was there any further evidence at this site of other faunal activity."

178 Botai in Kazakhstan: Alan K. Outram et al., "The Earliest Horse Harnessing and Milking," *Science* 323 (2009): 1332–35. DOI: 10.1126/science.1168594.

179 Holocene Galician art: Richard Bradley, *Rock Art and the Prehistory of Atlantic Europe: Signing the Land* (London and New York: Routledge, 1997).

180 the earliest known depiction of a horse roundup: Many of these depictions can be seen online. See, for example: www.rupestre.net/tracce/?p=577.

185 Bendrey's studies of goat domestication: Robin Bendrey, "From Wild Horses to Domestic Horses: A European Perspective," *World Archaeology* 44 (2012): 135–57, www.academia.edu/1785218/From_wild_horses_to_domestic_horses_a_European_perspective.

190 at least a thousand years earlier than the evidence at Botai: For an in-depth look at David W. Anthony's theories regarding early horsemanship and the spread of peoples and languages, read Anthony's *The Horse, the Wheel, and Language: How Bronze-Age Riders from the Eurasian Steppes Shaped the Modern World* (Princeton: Princeton University Press, 2007).

191 Horses thus created a new connectivity: There is a correlate from more modern times that shows how grass created the Mongol Empire, which spread during a particularly wet period. Check out Neil Pederson et al., "Pluvials, Droughts, the Mongol Empire, and Modern Mongolia," *Proceedings of the National Academy of Sciences* 111 (2014): 4375–79, www.pnas.org/content/early/2014/03/05/1318677111.abstract.

8. The Eye of the Horse

193 "The Lion and the Horse were arguing": The evolution of the eye caused Darwin even more psychic pain than did the evolution of horses. Gordon L. Walls's *The Vertebrate Eye and Its Adaptive Radiation*, published in 1942 (reprint edition, Eastford, CT: Martino Fine Books, 2013), attempted to explain the phenomenon of eye evolution and to lay out a framework for future research. Much has been accomplished since then, and scientists now do think they understand how such a complex organ could have evolved from astonishingly simple origins.

196 evolved a third color cone: Gerald H. Jacobs and Jeremy Nathans, "The Evolution of Primate Color Vision," *Scientific American* 300 (2009): 56–63. This excellent, in-depth but easily understood article by two of the world's experts in the field of vision discusses much more than the evolution of color vision. It explains the basics of how animals see and why color vision is an advantage in some—but not all—cases.

196 "face patches": For an in-depth discussion of neuroscience, color vision, art, and the recent discoveries of these intriguing areas in the brain, read the Nobel Prize–winner Eric R. Kandel's expansive but readable *The Age of Insight: The Quest to Understand the Unconscious in Art, Mind, and Brain, from Vienna 1900 to the Present* (New York: Random House, 2012).

197 The Canadian researchers Ian Whishaw and Emilyne Jankunis have found: Emilyne S. Jankunis and Ian Q. Whishaw, "Sucrose Bobs and Quinine Gapes: Horse (*Equus caballus*) Responses to Taste Support Phylogenetic Similarity in Taste Reactivity," *Behavioural Brain Research* 256 (2013): 284–90, www.ncbi.nlm.nih.gov/pubmed/23973764.

198 The biologist Christopher Kirk has compared: Amber N. Heard-Booth and E. Christopher Kirk, "The Influence of Maximum Running Speed on Eye Size: A Test of Leuckart's Law in Mammals," *The Anatomical Record* 295 (2012): 1053–62.

208 how do you figure something like that out for a horse: Brian Timney and Kathy Keil, "Visual Acuity in the Horse," *Vision Research* 32 (1992): 2289–93, www.psychology.uwo.ca/pdfs/SONA/articles/11-timney.pdf.

210 Ponzo illusion: Timney and Keil, "Horses Are Sensitive to Pictorial Depth Cues," *Perception* 25 (1996): 1121–28.

210 "Vision is not simply a window onto the world, but truly a creation of the brain": Kandel, *The Age of Insight*, 236.

212 horses are much better: In-depth, reader-friendly discussions of the science of equine vision are almost nonexistent. Michel-Antoine Leblanc's *The Mind of the Horse: An Introduction to Equine Cognition*, translated by Giselle Weiss (Cambridge, MA: Harvard University Press, 2014) provides a good science-based overview.

213 The Australian neuroscientist Alison Harman: Alison M. Harman et al., "Horse Vision and an Explanation for the Visual Behaviour Originally Explained by the 'Ramp Retina,'" *Equine Veterinary Journal* 31 (1999): 384–90.

9. The Dance of Communication

218 "Under a transport of Joy or of vivid Pleasure": That Darwin dared to compare the joy of children with the joy of horses could have raised a ruckus, but the public loved the book. And in this case, Darwin had firsthand knowledge, as his house was always filled with children and his pastures always filled with horses.

226 "Human beings are species-lonely": Thomas McGuane, *Some Horses: Essays* (Guilford, CT: Lyons Press, 1999).

230 The Oklahoma research psychologist Sherril Stone: Sherril M. Stone, "Human Facial Discrimination in Horses: Can They Tell Us Apart?" *Animal Cognition* 13 (2009): 51–61.

232 The French researcher Carol Sankey has shown: Carol Sankey has published a large body of research that examines many fundamentals of horse behavior, among them "Positive Interactions Lead to Lasting Positive Memories in Horses, *Equus caballus*," *Animal Behavior* 79 (2010): 869–75.

235 the German ethologist Konstanze Krüger has studied how horses learn: Konstanze Krüger et al., "The Effects of Age, Rank and Neophobia on Social Learning in Horses," *Animal Cognition* 17 (2014): 645–55.

236 Peacemaking among horses: For an excellent overview of Krüger's impressive body of research on the horse's social behavior, see "Social Ecology of Horses," in *The Ecology of Social Evolution*, eds. Judith Korb and Jürgen Heinz (Berlin: Springer, 2008), Krüger, available at http://epub.uni-regensburg.de/20253/1/Krueger_ecology_of_horse_behaviour.pdf.

237 The cognitive scientist Claudia Uller: Claudia Uller and Jennifer Lewis, "Horses (*Equus caballus*) Select the Greater of Two Quantities in Small Numerical Contrasts," *Animal Cognition* 12 (2009): 733–38.

238 when the dogs heard other dogs bark: Attila Andics et al., "Voice-Sensitive Regions in the Dog and Human Brain Are Revealed by Comparative fMRI," *Current Biology* 24 (2014): 574–78.

239 the British researcher Leanne Proops: Leanne Proops and Karen McComb, "Cross-Modal Individual Recognition in Domestic Horses (*Equus caballus*) Extends to Familiar Humans," *Proceedings of the Royal Society B* 279 (2012): 3131–38.

10. The Rewilding

243 "Only the wind": *Khadak* is a Mongolian film that can be seen for free on the Internet at www.imdb.com/title/tt0475241/.

251 "You should come to Mongolia next month": Most of the information in this chapter came directly from my visit to Mongolia and from conversations with scientists, government officials, and local people during that time. When I first began researching this book, finding background information on the Takhi and other horses in Mongolia was a challenging task. Though much has been written, most documents are either in the Mongolian language or in Russian. I came across two fantastic books by the American historian Jack Weatherford, which I gobbled up: *Genghis Khan and the Making of the Modern World* (New York: Crown, 2004) and *The Secret History of the Mongol Queens: How the Daughters of Genghis Khan Rescued His Empire* (Crown, 2010). These books are required reading for anyone interested not only in Mongolia, but in world history in general. Weatherford spent years living in and researching in Mongolia. Additionally, he was kind enough to spend several hours on the phone with me, helping me prepare for my own research trip. Substantive books about the history of Mongolia are difficult to find in the West, although this is slowly changing. To understand some of the nation's modern problems, try Manduhai Buyandelger's *Tragic Spirits: Shamanism, Memory, and Gender in Contemporary Mongolia* (Chicago: University of Chicago Press, 2013) and Tim Cope's accessible adventure tale *On the Trail of Genghis Kahn: An Epic Journey Through the Land of the Nomads* (Bloomsbury, 2013). Also available is *History of Mongolia: From World Power to Soviet Satellite* by Baabar (Cambridge, U.K.: White Horse Press, 1999). To learn more about the founding of Hustai National Park and the story of Jan and Inge Bouman, read *The Tale of the Przewalski's Horse: Coming Home to Mongolia* by Piet Wit and Inge Bouman (Zeist, the Netherlands: KNNV Publishing, 2006). This large volume has a myriad of information about the park itself and about the process of bringing the horses from the European zoos back to Mongolia. Included is a disk with an informative documentary that's well worth watching. Inge Bouman and her colleague Annette Groeneveld have also written a personal history booklet, privately published in 2008, about their experience, which Inge gave me during our lunch together: *The History and Background of the Reintroduction of the Przewalski Horses in Hustai National Park.* Since Hustai's founding, researchers from around the world have studied various aspects of the Mongolian ecosystem at the park. Some of these have been published in English, but are only available by visiting the park itself. Hustai has become an important center for bird conservation as well. Two local scientists, S. Gombobaatar and D. Usukhjargal, have published an excellent guide: *Birds of Hustai National Park* (2011). This, too, is locally published and available at the park's gift store.

ACKNOWLEDGMENTS

This book owes a great deal to a long list of people, including Deborah Cramer, author of *The Narrow Edge* and an exceedingly kind person. I'd also like to thank Joni Praded, whose calm, supportive suggestions were consistently both insightful and heartening, and Joan Chevalier, a wise counselor in the ways of the world.

Particular appreciation goes to Diane Davidson, Mark Spalding, and Angel Braestrup of the outstanding organization the Ocean Foundation, for their support of my interest in all things oceanic.

And to the many people who have read all or parts of the manuscript, including Matthew Mihlbachler of the American Museum of Natural History; Chris Norris of the Yale Peabody Museum of Natural History; Jason Ransom; Phyllis Preator; Kathleen Pratt (a knowledgeable reader and avid horsewoman) of the Cotuit Library; Hans Hofmann of the University of Texas at Austin; and Stephanie Kokal of the Horse-Tenders Mustang Foundation in Greenfield, New Hampshire.

And thanks to the many wonderful scientists and experts in the field of equine husbandry who hosted me in locations worldwide, including Piet Wit; Inge Bouman; park director Bandi Namkhai and the scientists and staff of Mongolia's Hustai National Park; Phyllis Preator and her friend Nettie Kelley, who drove me at length through their beloved Wyoming and filled me in on the recent history of the region's horses; thanks to Kim Scott and Eric Scott of the San Bernardino County Museum, who spent several days showing me local paleontological and geological sites and patiently answered my onslaught of questions during long, hot car rides; thanks to Stephan Schaal of the Senckenberg Research Institute and Natural History Museum of Frankfurt, Germany, who showed me the mille-feuille pages of the Messel research site; thanks to Isabelle Castenet, who spent hours talking to me about the horses on the cave walls discovered long ago by her great-grandfather; thanks to Herwig Radnetter for introducing us to his three magical white stallions.

And a very special thanks to Laura Lagos of Galicia and her colleague Felipe Bárcena Varela de Limia of the Instituto de Investigación y Análisis Alimentarios, Universidad de Santiago de Compostela, for organizing one of the most fascinating research trips I've ever experienced. Together they made sure that I saw so many different areas of Galicia where the Garranos roamed, explained the deep history of their beloved countryside, and provided wonderful meals with the best of local wines. Thanks to them for introducing me to Javier Álvarez Blázquez (Asociación de Gandeiros de Cabalos do Monte da Groba), Xosé Lois Vilar (Instituto de Estudos Miñoranos and S.O.S. Groba), Xilberte Manso de la Torre (Instituto de Estudos Miñoranos and S.O.S. Groba), Modesto Domínguez Roda (Asociación de Gandeiros de Cabalos do Monte da Groba), José Manuel Rey García (Parque Arqueológico de Campo Lameiro), Dr. Jaime Fagúndez (heathers expert and professor in the University of A Coruña), Dr. Roberto Hermida, Dr. Santiago Bas.

And thanks to the many, many scientists who kindly forwarded me research studies and spoke at length with me—often more than once—including Harald Floss, John Turner, Gus Cothran, Philip McLoughlin, Darrin Pagnac, Christine Janis, Philip Gingerich, Ken Rose, Chris Beard, Martin Fischer, Shari Ackerman-Morris, Claudia Uller, E. Christopher Kirk, Elise Renders, Neil Pederson, Sarah Feakins, Jeffrey Stevens, Caroline Galloway Strang, Kevin Uno, Nick Famoso, Lee Olynyk, Grant Zazula, Dale Guthrie, Robert Raynolds, Brian Kooyman and Len Hills, Stuart Fiedel, Gary Haynes, John Tyler Bonner, Craig Packer, John Hoffecker, Sebastián Jurado Piqueras, Robin Bendrey, the Kokal family, David Anthony and Dorcas Brown, Robert Cook, Sandra Wise, Margery Coombs, Joseph Carroll, Gerald Jacobs, Brian Timney, Karen Murdoch and Lukas, Sherril Stone, Konstanze Krüger, Nicole Waguespack, John Wible, Gregory P. Wilson, Gina Semprebon, Timothy J. Gaudin, Zhe-Xi Luo, Jacquelyn Gill, Harry Jerison, Andrew Hill, Lillian Spencer, Pamela S. Soltis, Matthew Sisk, Ross Secord, Megan Nordquist, Mike Voorhies, Thomas Barfield, Paula DePriest, Julien Riel-Salvatore, Melissa Songer, Donald Prothero, Budhan Pukazhenthi, William Fitzhugh, Brianna McHorse, Robert W. Meredith, Bolortsetseg Minjin, Karyn Malinowski, Katherine Albro Houpt, Caroline Strang, Lynne Isbell, Dennis Jenkins, Tom Tobin, Gregory Curtis, Anthony Fiorillo, Richard Stucky, Roland Kays, Christopher Hemmings, Lawrence Straus, David Archibald, Lee Boyd, Elizabeth Kellogg, David Grossnickle, Sandra Engels, J. M. Adovasio, Richard B. Alley, Ray Bernor, Luke Holbrook, Melinda Zeder.

To my tolerant family—Kay, Susan, Diana, Bruce, Judy, Mike, and Greg—love and appreciation from the bottom of my heart. Writing a book like this is an all-out effort. Without their patience, the task might not have been completed.

And thanks of course to my supportive agents, Wendy Strothman and Lauren MacLeod, and to my editor, Amanda Moon; to Melissa Cavill of the Cotuit Library, who never told me that any book I wanted to read, no matter how arcane, was too difficult to find (what would the world be without librarians?); to Laird Gallagher, for his phenomenal efforts in assembling the art for this book; and to Annie Gottlieb, to whom I am so deeply indebted.

INDEX

Page numbers in *italics* refer to illustrations.